自 然 之 子

——美国当代风景园林大师理查德·哈格作品集

[意] 卢卡·玛利亚·弗朗西斯科·法布里斯 著

戴菲 李保峰 译

中国建筑工业出版社

著作权合同登记图字：01-2011-5674号

图书在版编目（CIP）数据

自然之子——美国当代风景园林大师理查德·哈格作品集/（意）法布里斯著；戴菲等译．—北京：中国建筑工业出版社，2012.10
ISBN 978-7-112-14192-0

Ⅰ．①自… Ⅱ．①法…②戴… Ⅲ．①景观设计-作品集-美国-现代 Ⅳ．①TU986.2

中国版本图书馆CIP数据核字（2012）第055930号

© Copyright 2010 by Maggioli S.p.A.
All rights reserved.

Nature as Lover by Luca Maria Francesco Fabris

本书由意大利Maggioli Editore出版社授权翻译出版

责任编辑：程素荣
责任设计：赵明霞
责任校对：陈晶晶　王雪竹

自然之子
——美国当代风景园林大师理查德·哈格作品集
[意]卢卡·玛利亚·弗朗西斯科·法布里斯　著
戴　菲　李保峰　译
*
中国建筑工业出版社出版、发行(北京西郊百万庄)
各地新华书店、建筑书店经销
北京嘉泰利德公司制版
北京方嘉彩色印刷有限责任公司
*
开本：880×1230毫米　1/16　印张：10$\frac{1}{2}$　字数：300千字
2013年1月第一版　2013年1月第一次印刷
定价：**88.00**元
ISBN 978-7-112-14192-0
（22260）

版权所有　翻印必究
如有印装质量问题，可寄本社退换
（邮政编码 100037）

目录
Contents

序言   4
理查德·哈格：自然之子
卢卡·玛利亚·弗朗西斯科·法布里斯

设计项目   17

1. 西雅图煤气厂公园，华盛顿州   18
2. 伯克利北部滨水公园，加利福尼亚州   36
3. 案例研究之家   40
4. 杜克斯办公室   42
5. 罗斯福纪念碑竞赛   44
6. 弗莱艺术博物馆   48
7. 巴特尔纪念馆   56
8. 美洲园艺博览会竞赛   60
9. 乔丹公园／埃弗里特码头公园   64
10. 加利福尼亚大学伯克利分校林间纪念广场   68
11. 布洛德尔保护地   70
12. 斯塔米诊所   94
13. 布鲁姆住宅   98
14. 西雅图中心   102
15. 拉维莱特公园竞赛   106
16. 吉尔曼村庄／吉尔曼林荫大道   108
17. 亨利·M·杰克逊联邦大厦   112
18. 梅里尔庭院式联排住宅   118
19. 华盛顿州生态总部设计竞赛   124
20. 美国驻葡萄牙里斯本的大使馆和领事馆   126
21. 维克托·施泰因布吕克公园   128
22. 狄龙住宅   132
23. 蓝绿公园竞赛   138
24. 库克斯住宅   142
25. 萨默维尔住宅   144
26. 泰特·弗里斯住宅   150

理查德·哈格自传   157

附录   163

序言
Foreword

理查德·哈格：自然之子
卢卡·玛利亚·弗朗西斯科·法布里斯

  理查德·哈格无论是作为一个人，还是作为一位设计师的优势都在于他独立思考的能力。他笑容爽朗，思维开阔。他的眼中总闪烁着探索的光芒，折射出智慧的广度与深度。也许，细节是捕捉这位景观设计巨匠个性的最好方式。当他第一次带我漫步在他最著名的作品西雅图煤气厂公园时，那一刻的安静胜过任何华丽的辞藻与讲述，所谓沉默是金不过如此。在参观的过程中，哈格似乎更着迷于让我了解公园本身而非他作为一位知名教授和建筑师的常用模式以及手法。我恍然反应过来，自己竟有机会去领悟一个项目或者说有效的实践经验的精髓所在，而不是要听取对某种方法的描述。此时此刻，设计带来的体验传递了设计理论。如果我以一种超越视觉的感官去感知这个公园，我便可以理解事物背后的精髓，并且由此抽象出产生这件作品背后的自然哲学。

  它多大程度上属于这位来自肯塔基州的年轻人天性的一部分？其父亲已经是一名园艺家和植物杂交技术的开发者。的确，哈格对自然的领悟随着年龄而成熟，通过自身的尝试去理解自然界的韵律与特性。然而，我确信这一份对自然界生生不息、循环往复的敬畏与尊重，是这位年轻人与生俱来的天性；而且，第二次世界大战后，

Richard Haag: A Lover of Nature
Luca Maria Francesco Fabris

Richard Haag's strength as both a man and a designer lies in his independence of thought. His wide smile is as open as his enquiring mind; his eyes, always a-gleam with the light of discovery, reflect an intelligence that is both magnanimous and perceptive. Perhaps anaecdote is the best way of encapsulating the personality of this giant of landscape architecture. The first time he took me round his most famous work – the Gas Works Park in Seattle – silence prevailed over words and explanations. Throughout the walk, Haag seemed much more interested in the research that had led me to get to know his work than in talking about his own modus operandi as a famous teacher and architect. It took me a little time to see that I was being given not an opportunity to hear a method described but rather the chance to absorb the essence of a project and of the manner in which it had been put into effect. Experience here conveyed theory. If I looked beyond the mere visible, I could understand the mechanics behind things, and from that I could then 'abstract' the natural metaphysics which inspired such work.

How much of this was already part of the young kid from Kentucky, whose father had been a plant nursery owner and developer of plant hybrids? True, Haag's relationship with Nature must have matured over the years, through his own attempts to understand the rhythms and characteristics of the natural world. However, I am sure that this reverence and respect for the cycle of Nature was something that the young man was born with; the long period of study he spent in – a

哈格在日本很长一段时间略带神秘色彩的学习和研究开发了他天性中的这一面。

理查德·哈格对日本景观艺术的独到理解与诠释是其设计作品的关键特色之一。有的时候这种影响表现得非常明显，而有的时候又显得含蓄。西雅图煤气厂公园和布洛德尔保护区作为他最具盛名的两件作品，是这种特定的景观艺术如何被移植的完美案例。这些项目中的第一个项目坐落于伸出尤宁湖的半岛上，由巨大的延伸的草地构成，略微起伏错落的地形完美安置着油库光秃秃的机器设备，就像在京都龙安寺禅的庭园里一样，象征海洋的苔藓上构建着沉寂的石头。同样，在华盛顿州布洛德尔保护地，比起现在不幸遭到破坏的水平庭园，苔藓庭园更能例证他早期设计庭园的自然性。死（慢慢腐朽的树桩）在这里与生（不可胜数的贴近潮湿苔藓之中的花朵）共存。这种并列的设计手法将显然无序的景观与东方园林的经典场景完美地糅合了。

哈格在他的每一次设计中，都通过重新定义连续性来调停边界。煤气厂公园的边界是前方的地平线而非包围着场地的水岸线，西雅图时常变化的城市天际线也融入进来，成为整体设计的一个组成部分。然而，在布兰布里奇岛上，无明显边界的布洛德尔保护地则将人们的视线引向场地内部，恰好反映了人与大自然的现存关系。

作为唯一的一位两度获得最高设计奖

still somewhat mysterious – Japan soon after the Second world War would simply develop this innate aspect of his character.

His own personal interpretation of the Japanese art of landscape is one of the key features of Richard Haag's work. Sometimes the influence is explicit, sometimes it remains more concealed. The two greatest achievements of his oeuvre – the Gas Works Park and the Bloedel Reserve – are perfect examples of how that specific art of landscape can be transposed. Located on a peninsula that juts into Union Lake, the first of these projects comprises a massive expanse of grass, gentle rising and falling to become the perfect setting for the bare machinery of a gas plant – just as the moss (a symbol of the ocean) frames the silent stones in the Zen garden of the Ryoanji Temple (Kyoto). And in the Bloedel Reserve, more than the Garden of Planes (now unfortunately disfigured), the Moss Garden exemplifies the naturalness of the very earliest gardens; death (slowly decaying tree trunks) here coexists with life (innumerable flowers nestling amidst the damp moss), and this juxtaposition links the apparently disordered landscape with the classical image of the Eastern garden.

In each case, Haag intervenes through redefining continuity. The boundary of the Gas Works Park is not the water around it but the horizon beyond; the ever-changing skyline of Seattle itself becomes part of the whole design. And on Bainbridge Island, the apparently limitless Bloedel Reserve obliges one to look inwards, to reflect upon the present relation between humankind and Nature.

The only person to have twice received the President's Award for Design Excellence –

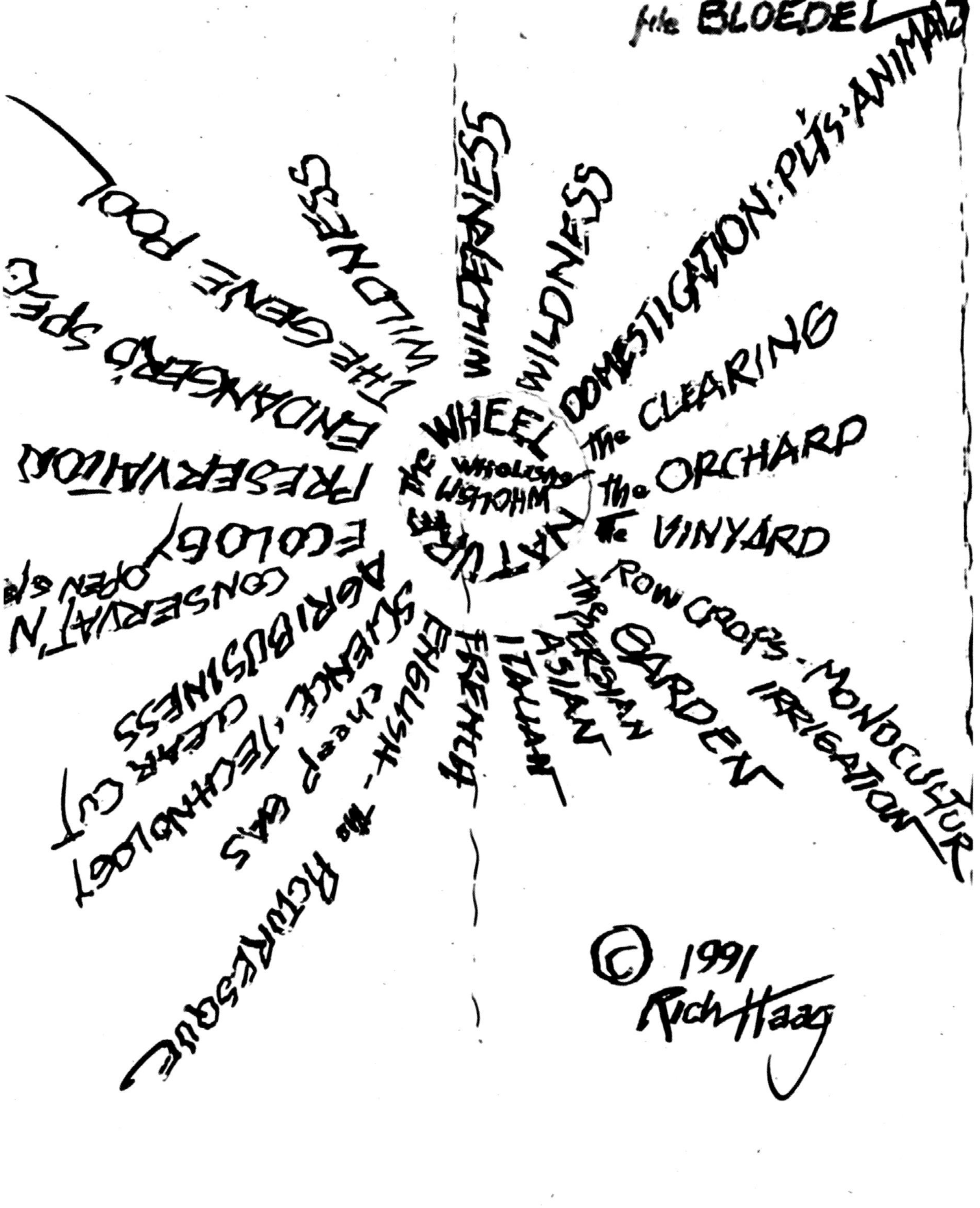

file BLOEDEL

The Wheel of Wholism

Central hub: **the WHEEL of WHOLISM** / **the WHOLE of NATURE**

Spokes (clockwise from top):
- WILDNESS / WILDNESS / WILDERNESS
- DOMESTICATION: PL'TS · ANIMALS
- the CLEARING
- the ORCHARD
- the VINYARD
- Row Crops — MONOCULTUR / IRRIGATION
- GARDEN — Persian, Asian, Italian, French, English
- the PICTURESQUE
- cheap GAS / SCIENCE / TECHNOLOGY / CLEAR CUT
- AGRIBUSINESS
- CONSERVAT'N / OPEN S/A
- ECOLOGY
- PRESERVATION
- ENDANGERD SPEC / the GENE POOL

© 1991 Rich Haag

（美国风景园林协会最高荣誉）的景观设计师，理查德·哈格自1958年以来一直在西雅图的华盛顿大学任教，并于1964年创办了该校的风景园林系。自从毕业于哈佛大学之后，他的事业从加利福尼亚州开始起步，并且在专业活动中不断揭示设计与教学这两者间的相互作用。这种动态渗透的结果不仅仅是建立了培养风景园林师的学校，一个名副其实的团体，而且也表现在帮助形成俯瞰普吉特海湾的"翡翠城"恰如其分面貌的一系列项目之中。

公共空间在哈格的设计中被定义为生活的空间。随着植物和矿物质参与到阳光、色彩和表面的对话中，这些空间成为城市生活中不可或缺的一部分。那些有冲击力的构造常常能让景观本身变成一种建筑。在每一个项目中，被塑造的空间汲取时间的韵律——生长、开花、休眠——转化为自身的节奏。它意味着每个项目经历时间而成熟，没有这个进程，每个细节背后的设计意图将会丧失而变得模糊。哈格的设计几乎总是使用乡土植物。这样，他的作品创造了一种由美国西北海岸乡土植物定义的新的景观语言。就像在东方的庭园设计艺术中，创新性和稀缺性已经不再是最重要的方面了，取而代之的是设计中的整体布局。确实，设计时注意力都集中在植物的作用上，但是哈格从未忽略空间的实用性，使其必须满足使用者的需求。

the highest accolade of the American Association of Landscape Architects – Richard Haag was in 1964 also the founder of the Landscape Architecture Department at the University of Washington in Seattle, where he had been working since 1958. Ever since his own graduation from Harvard, his own career – which began in California – has revealed the interaction of these two aspects of his professional activity. And the result of this dynamic osmosis has been not only the establishment of a School – a veritable community – of 'landscapers', but also a range of projects that have helped to define the very appearance of the 'Emerald City' overlooking Puget Sound.

Public spaces as designed by Haag are spaces to be lived in; they are veritable parts of city life, with the vegetable and the mineral interacting in a dialogue of light, colour and surface. Very often, powerful tectonics serve to make the landscape itself into architecture. In every project, the space created absorbs the temporal rhythms of Nature – of growth, flowering and repose – to make them its own. This means that each project withstands time by maturing with it, without this process leading to the loss or blurring of the intention behind each individual detail. Haag's designs almost always use local varieties of plant. Thus, his work has created a new language of landscape as defined by the species of vegetation indigenous to the north-west Pacific coast of the USA. In his plant nursery ay Everett (just to the north of Seattle), he breeds the shrubs and trees that are now recognised as his signature. Just as in the Oriental art of garden design, it is not originality or rarity that count but rather the

在他的私家庭园设计中，哈格通过让场地响应"干扰"来诠释场地的本身，植物与建筑混合在一起塑造场景，从而表达出庭园主人的性格。这里的空间或许是有限的，但这些设计都反映出理查德·哈格多年来总结出的理论与实践经验。

"哈格模式"可以从他西雅图小工作室钉在墙壁上的打印纸张中找到。它表达得那么明显，以至于都让人有点疑惑。然而，就像所有的明显事物一样，这种模式很难被应用于实践当中，尤其是在设计领域。人类偏偏擅长将简单的事物复杂化，从而加快消耗创造世界的能量，致使人工化的痕迹越来越重。在他的模式中，哈格选择了正确的方向去设计，仅仅依据六个要素：空间、尺度、循环、大地、水和植物，结合了大自然产生的力量与人类活动的主观意愿。这类活动常被视为短视而不能纵观全局的，所以迷失在华丽的细节中。

哈格的景观设计常可以等同于环境规划，不仅仅是因为他把人类科学技术融入环境当中，更因为他预示了一种自然与技术间的平衡。没有什么比再现自然的鬼斧神工更难的了。胜败往往只差一步之遥，实验是前进的唯一途径。将实验成果转换为实践经验然后再提炼成为一种行为模式，需要花费大量的时间。以哈格自己为例，他是第一位进行生物修复实验的景观设计师，运用自然的技术对被化学污染的场地进行改造，以正视后工业基地作为城市公

layout of the design as a whole. True, emphasis is placed upon the role of plant life; but Haag never loses sight of the fact that the space must be functional, must meet the needs of those who use it.

In his private garden designs, Haag interprets the site itself by allowing it to respond to 'interference'; the plants are used to create settings that blend with architecture at the same time as they reflect the personality of the garden's owner. Space may be limited here, but these designs reflect the theory and practice that Haag has developed over the years.

The 'Haag formula' is to be found on a printed sheet that is pinned to the wall in his small studio in Seattle. What it says seems so obvious that one is left a little perplexed. However, like all obvious things, the 'formula' is very difficult to put into practice, especially in the field of design; for as it becomes increasingly 'cultivated', humankind seems to specialise in complicating things, in increasing the entropy of the created world. For his part, Haag encapsulates the correct approach to design in just six words: 'Space – Scale – Circulation – Earth – Water – Plants', combining the generative power of Nature with the volitional nature of all human activity (an activity that all too often proves short-sightedly incapable of seeing the whole, and so loses itself in the dangerous magnificence of details).

Haag's landscape architecture has always been the equivalent of environmental planning – not simply because it involves the application of human technologies to the environment but because it is predicated upon a balance between Nature and technology. Nothing is more difficult than recreating the alchemy of the created world. Failure is often

the Cosmos is an experiment
the UNIVERSE is a park
the EARTH is a pleasure ground
NATURE is the theater
the LANDSCAPE is our stage

Let us write the script
direct the play
and embrace the audience
with compassion and joy
for LIFE

Richtaag '85

共空间的再开发。这一切都起源于1969年在西雅图建造的油库公园，从那以后的每个项目，都是这第一步迈出之后的结果与续篇。

哈格伏案设计的工作桌便是所谓"有组织混乱"的最好例证。桌子四周贴满了色彩明快的便笺纸，上面满是用铅笔摘要记载的单词和灵感。简短的记录精准地反映出哈格作为一个设计师抑或是艺术家的思想。其中，我印象最深刻的莫过于来自伏尔泰的名言："不要让完美成为优秀的敌人"。这是帮助人们理解这位伟大的美国风景园林师作品的另一线索。

这么多年来，理查德·哈格作为一名风景园林师，一直从事工作和教学，从未出版反映自己理论与实践思想的作品集。但是，他的设计理念已经在无数次的热情洋溢又慷慨激昂的演讲和报告中被全世界所知晓。同样，他从来没有把自己的项目和设计作品收录到一个独立的作品集中，也不曾想让别人这么做。这本图文并茂的专著，在世界上首次回顾了理查德·哈格合伙人公司发展历程，无论是在美国还在世界各地，对把景观建筑作为环境规划工作关键性考量的人们，不仅产生了而且还将继续产生巨大的影响。作为一名欧洲的研究者与建筑师，我非常荣幸撰写这部专著。这是个相当大的责任和巨大满足感的源泉。

在哈格合伙人公司的大量项目中，本

just a short step away, with experimentation the only way forward; and it takes time – a lot of time – to transform that experimentation into experience and thence into a model for activity. Haag, for example, was the first landscape architect to experiment with bio-remediation – the use of natural techniques for the reclamation of chemically polluted terrain – and to envisage the redevelopment of post-industrial sites as public urban spaces. All of this began with the Gas Works Park (Seattle) in 1969, and everything thereafter is a consequence of those first steps.

Haag works at a drawing-table that is the perfect example of organised chaos. All around it are brightly coloured post-its with pencil jottings of words and ideas – short notes that give a very precise idea of the man, the designer and the artist. One among the many that struck me was this warning by Voltaire: "Don't let the perfect become the enemy of the good" – another clue that helps one to understand the work of this great American landscaper.

In all the years that he has been teaching and working as a landscape architect, Richard Haag has never published a collection of writings on his ideas regarding theory and practice - ideas which he has, however, discussed with great enthusiasm and generosity in numerous lectures throughout the world. Similarly, he has never gathered together all of his projects and designs in a single corpus; nor has he wanted anyone else to do so. Combining texts and images, this monograph is the first in the world to recount the history of a studio – Richard Haag Associates – that has and continues to have such influence not only within the United States but wherever

书选取了50多年间的26件作品。这半个世纪的历史通过大量首次被发表的照片和图像进行回顾,这些大量的、通过时间标记清晰的插图,原本是用作工作档案和研究资料,没有打算发表。在华盛顿大学西雅图校区专业收藏图书馆里的私人档案现在被公开,将在非常专业的工作中变成使用工具:景观建筑,就像理查德·哈格注释的那样,是抱着愉悦感和坚定的信心,拥抱自然作为情人的唯一职业。

2010年7月于米兰

people have taken landscape architecture as a key point of reference in their approach to environmental planning. And the honour of producing this monograph has fallen to me, a European researcher and architect. It is a sizeable responsibility - and an enormous source of satisfaction.

Amongst the Haag Associates' numerous projects, this book covers 26 from a period spanning more than 50 years. This half-century of history is recounted using images and photographs that are often being published for the first time; clearly marked by the passage of time, numerous of these illustrations were originally produced as working documents, as studies that were never intended for publication. A private archive, now at University of Washington, Seattle, Special Collections Library, is thus being made public, to become an instrument for use in a very special kind of work: landscape architecture, which Richard Haag notes, with both amusement and unshakeable confidence, "is the only profession that embraces Nature as a lover".

Milan, July 2010

设计项目
Project Designs

1.

西雅图煤气厂公园，华盛顿州

北纬：47°38′42.98″

西经：122°20′6.00″

地点：西雅图，诺斯莱克路 1801 号
时间：1969 ~ 1975 年，持续
状况：已建成。新工事：修复儿童游戏场、安装隐蔽摄像头、塔周围栅栏有待拆除
项目设计：奥尔森·沃克及合作者；迈克尔·G·安斯雷
项目经理：理查德·哈格
设计团队：理查德·哈格事务所（劳瑞·奥林，道格拉斯·图马，斯蒂芬·G·雷，健日·中野）
顾问：Richard Brooks (bioremediation)
项目委托方：Seattle Department of Parks and Recreation
工程设计：Arnold, Arnold and Associates (strutturestructural); Beverly A. Travis and Associates (elettricoelectrical); Miskimen Associates (impiantisticamechanical)
主要承建方：Bordner Construction Company; Davis Court Construction Company; George Adams
制图方：Laurie Olin, RHA
摄影师：Mary Randlet, Luca M.F. Fabris, Patrick Waddell, Richard Haag Associates
艺术家：Sundial: Charles Greening, Kim Lazare
场地面积：8 公顷（20 英亩）
建造费用：200 万美元

西雅图煤气厂公园对公众开放 35 年以来，在世界范围内被广泛地认为是废弃的工业地转换为城市公园的一种模式。它是一连串项目中的首个，逐渐改变了公共城市空间被设计、被解读、被使用的方式。成效随着时间的流逝得以突显，哈格项目的前瞻性体现在整体的工作方式中，为了发展非常规技术以解决场地的地面污染问题，它第一次把广泛的不同专业技术人士聚集在一起。如今，整个方案可能被贴上整体的、跨学科的标签，但是在当时，那是一个勇敢的研究者在该领域努力的成果。这种方法的直接采用在今天随处可见，比如：彼得·拉茨设计的杜伊斯堡景观公园（1995 年）和麦克格雷戈及其合伙人公司设计的悉尼英国石油公司遗址公园（2006 年），都展现了近年来美学视角的激烈变革。崇高和美丽的地方不再是人们苦苦追寻的，而是与我们每天城市生活的日常喧闹和谐共存在一起。

为充分了解西雅图这片区域发生事情的重要性，应该回顾一下煤气厂公园的历史。这个项目遭遇到一系列的问题，并以令人惊讶的具有划时代意义的方案来解决。当巴里·布衣发表开创性的环保读物《封闭的循环》的时候，该方案展现了对环境警觉的第一道光芒。这种环境意识最终导致西雅图成为美

Open to the public thirty five ago, the Gas Works Park in Seattle is considered worldwide as a model for one type of urban-park reconversion of disused industrial areas. It was the first in a chain of projects that have gradually changed the way public urban spaces are designed, interpreted and used. Its validity borne out by the passage of time, Haag's project was predicated upon an overall approach that, for the first time, brought together a range of different professional expertise in order to develop non-conventional techniques for the resolution of the self-evident problem posed by such sites: ground pollution. Nowadays the whole scheme might be labelled as a holistic and 'trans-disciplinary' one, but at the time it was the fruit of the efforts made by one courageous researcher in the field. Direct descendants of the approach adopted here can be seen, for example, in Peter Latz's Duisburg Landschaftspark (1995) and the BP Park in Sydney by McGregor + Partner (2006), each of which reveals a radical change in aesthetic perceptions in recent years. The sublime and the beautiful are no longer where one might expect to find them; they now co-exist quite happily with the normal chaos of our everyday urban existence.

To understand the full importance of what was done in this area of Seattle, one should look back over the history of the Gas Works Park, given that the project embraces a whole series of problems and solutions which are still of surprising contemporary relevance. Created in the years when Barry Commoner published that seminal environmentalist text *The Closing Circle*, the scheme also reveals the first glimmerings of that environmental awareness which would ultimately lead to Seattle becoming one of the USA's capitals of research into sustainable development.

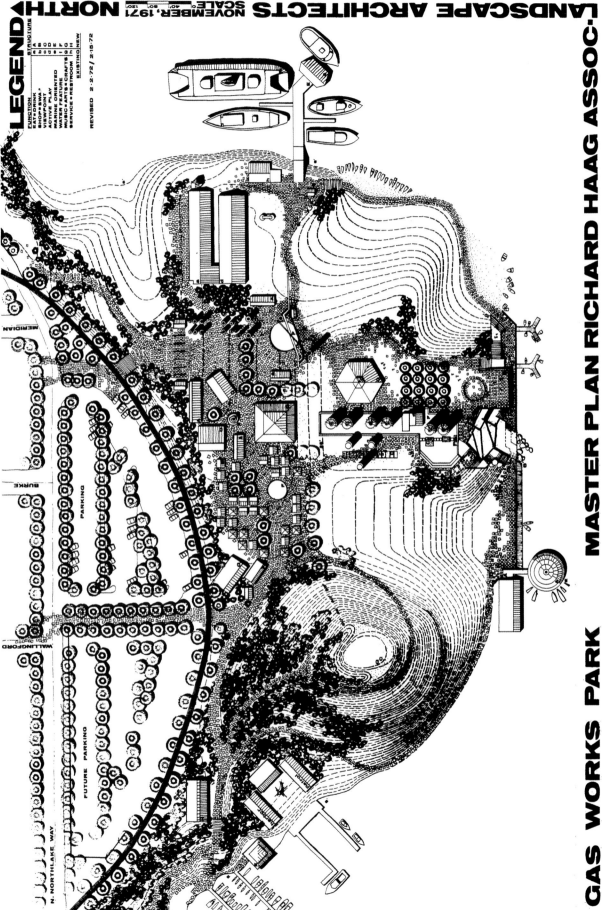

国研究可持续发展的中心之一。

煤气厂于 1906 年首次开业，它坐落于西雅图城对面一个叫布朗点的海角，海角向南倾斜于尤宁湖。当 1956 年煤气厂倒闭的时候，市政府决定购买这块地，并主张将其复兴为开放公园。实际上，奥姆斯特德 1904 年就建议在这块被树林覆盖的土地上建造公园。然而，时间流逝事情却没有任何进展，因为在老旧的工业厂区建造什么和如何解决场地污染的问题并不清楚。在 20 世纪 60 年代中期，理查德·哈格和越来越多的人开始争论，在这个公园项目中保留工业设施，以挽救这个城市历史肌理的一部分。他的项目（1969 年）于 1971 年被批准，但是很快就出现了争议，因为要移除 8.2 公顷被煤气生产过程中副产品污染的土壤需要高昂的成本，这些副产品包括苯、萘、甲苯和其他产生有毒气味的多核物质。

要建造公园的海角，实际状况是"完全没有无污染的土壤和自然的植物生命；到处是地下管道，其中装满了工业废渣和有毒碳氢化合物。70% 的表面是密不透风的，由混凝土、沥青、焦油和其他惰性材料压实而成。其他的地面区域，特别是环绕场地周边的区域，可能没被压实，但是乱扔着混凝土废渣、碎石、橡胶件、金属碎片、玻璃、木材等，没有被覆盖的地方几乎找不到。最严重的污染区域是在紧靠贮气罐的附近。任何植被在这里生长几乎都是不可能的，还必须采取谨慎措施以防止污染扩散到周边地方。"（理查德·哈格在 S. 威姆斯，1980 年）。和化学工

Located opposite Seattle city on a promontory once known as Brown's Point, which slopes southwards towards Lake Union, the gas works were first opened in 1906. When they closed down in 1956, the City Council decided to buy the site and revive the notion of turning the area into a public park; the Olmsteds had, in fact, suggested the creation of a park in 1904 when the site was forested. However, years passed with no further progress being made, because it was unclear what should be done with the old industrial plant and how the problem of site pollution was to be resolved. In the mid-1960s, Richard Haag – together with a growing number of people – began to argue for a park project that would maintain the industrial structures, thus saving a part of the historical fabric of the city. His project (1969) was approved in 1971, but immediately raised controversy because of the high costs of removing some 20.5 acres of soil contaminated by the products of the gas-making process: benzene, naphthalene, toluene and other polynucleates that generated noxious odours.

The promontory on which the park was to be created was, in effect, by then "... totally without unpolluted groundsoil and natural plant life; riddled with underground tubes and pipes, it was a 'Cinderella' rich in industrial residues and toxic hydrocarbons […] 70% of the surface was impenetrable, being highly compacted by the use of such consolidating substances as concrete, asphalts, tar and other inert materials. The other ground area, particularly around the perimeter of the site, may not have been compacted but was littered with concrete waste, rubble, pieces of rubber, fragments of metal, glass, wood, etc, so that very little of the ground area was left uncovered. The most heavily polluted terrain was that in the immediate vicinity of the gasometers. It was unlikely that one could get any type of vegetation to grow there, and care had to be taken to prevent the contamination spreading to the terrain nearby..." (R. Haag in S. Weems, 1980). Working together with the chemical engineer Richard Brooks, Haag came up with the solution of removing

程师理查德·布鲁克斯一起工作，哈格找出了从地表除去污染物的办法，即通过翻耕锯末、灌溉污水淤泥和其他有机物来激活土生土长的微生物。这是生物修复的第一个案例，这个过程意味着在相对较短的时间内，土地能够被开垦。土地上没有接受这种治疗的部分用作假山的基地，提供一个站在煤气厂上俯瞰海湾、尤宁湖和西雅图市中心的场所。所有这些地方都覆盖了一层薄薄的引入土壤，土壤中混合着像堆肥、碎草这样的其他材料。

公园于1975年对外开放，特点是覆盖着多孔铺地或附近割下的草的动态模式。在公园的北部有一排常青树作为屏障，沿着老工业铁路轨道种植（隐蔽着一个小停车场）。景观的整体布局突显了煤气厂巨大垂直结构的影响，就像一个禅宗布局，整体感觉是留给置身其中的人们自由地解读。海拔、视野和视角，曲线和路径的变化——都形成了一个拥抱着工业遗迹、简单而平衡的叙述。取消或者修改任一个特征都将改变构成整体表达的含意与诗情。整个项目被深思熟虑地轻描淡写，使它能完美地应对随着时间不断发展的生活方式和休闲活动的变化。

the contaminants from the surface by activating the indigenous bacteria by tilling sawdust, sewage sludge and other organic compounds. One the first examples of bio-remediation, this process meant that, in a relatively short time, areas of the land could be reclaimed. Various portions of the earth that did not undergo this treatment were used as the base for the artificial hill that stands over the Gas Works, affording a view over the bay, Lake Union and down-town Seattle. All of these areas where then covered with a thin layer of imported earth mixed together with other materials such as compost and chopped-up grass.

Opened in 1975, the park is characterised by the dynamic modeling of terrain, which is covered with either porous paving or closely-mown grass. There is a curtain of evergreen trees to the north of the site, following the curve of the old industrial railway track (and concealing a small car park). The whole layout of the landscape is designed to highlight the impact of the enormous vertical structures of the Gas Works; just as in a Zen layout, the reading of whole is left to the free interpretation of the individual within the space. The changes in elevation, the views and prospects, the curves and footpaths – all form one single and balanced narrative embracing the remains of the industrial plant. The removal or modification of one feature would alter the covert significance, the poetry, of the composition as a whole. The entire project is deliberately understated, making it perfectly capable of responding to changes in lifestyle and leisure activities which may develop over time.

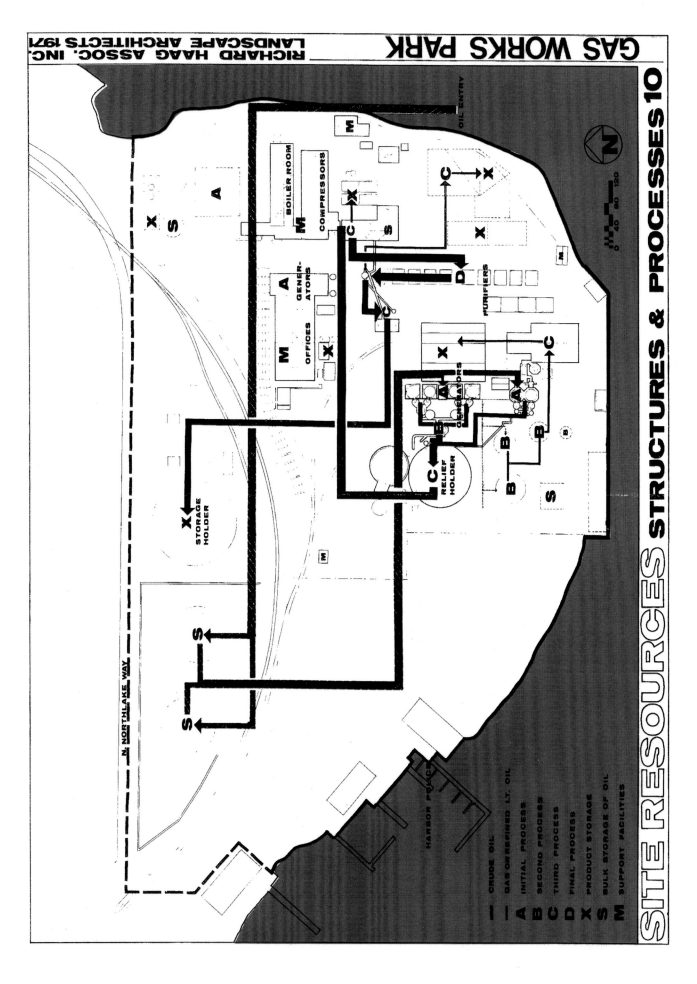

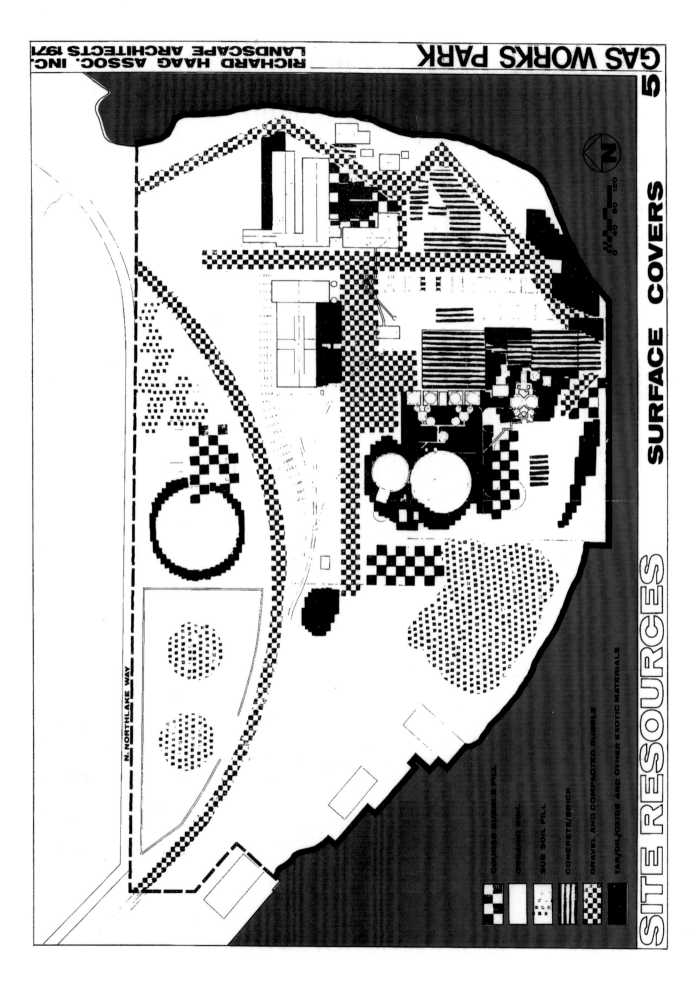

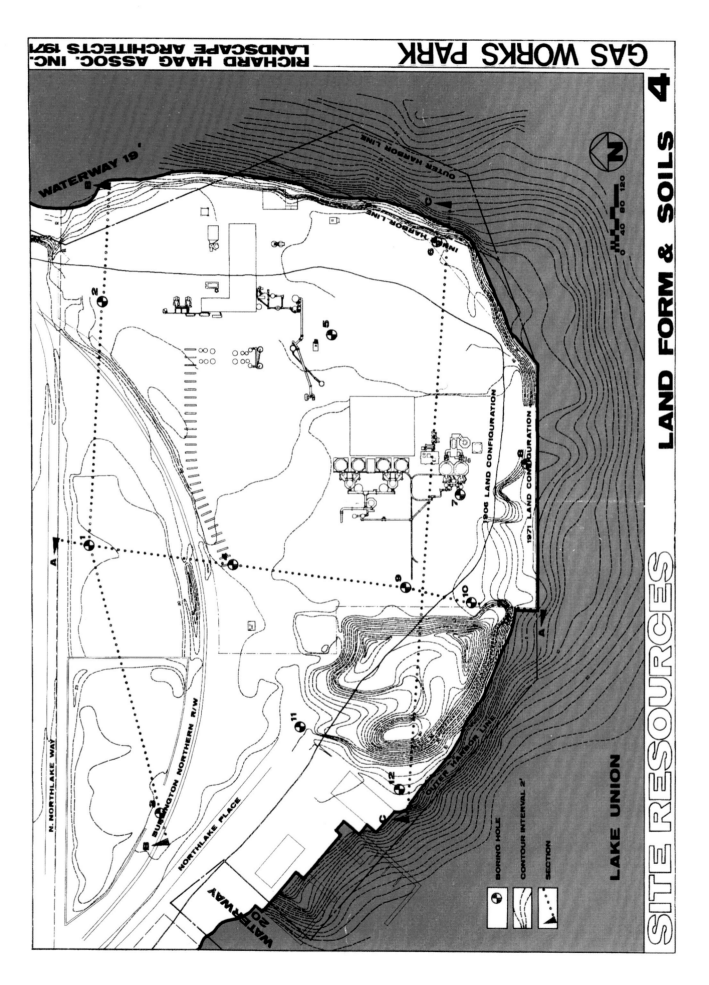

2.

伯克利北部滨水公园，加利福尼亚州

北纬：37°52′20.32″
西经：122°19′8.45″

地点：旧金山湾，伯克利西部，加利福尼亚州
时间：1987 年
状况：未建成
项目经理：约翰·诺斯莫尔·罗伯茨，景观设计师
设计团队：理查德·哈格事务所，约翰·诺斯莫尔·罗伯茨，艾格尼丝·德纳斯，环境艺术家
项目委托方：City of Berkeley
工程设计：Dillingham Associates; Harding Lawson Associates; Moore Iacofano Goltsman; Willard Bascom (dinamiche costiere shoreline dynamics); Richard Brooks (consulente bonifica reclamation consultant)
制图方：Richard Haag
摄影师：Mary Randlett, RHA
场地面积：39 公顷（97 英亩）

如同在美国大多数沿海城市一样，伯克利的滨水区最初并非房屋遍布，而是连接港口和城市腹地的商业和工业活动开发了这块土地的价值。随着时间的推移，工业的发展产生了一条密不透风的基础设施带，将城市和大海完全隔绝了开来。有些时候，工业活动甚至延伸过海岸线，占据了由生活和工业废弃物形成的人工填海区。

1987 年，理查德·哈格提出这个在美国大学城里位于码头附近的 97 英亩广阔圩田，应该被改造为一个公共公园，从而将露天垃圾堆变成一个新的城市"绿肺"。受到西雅图煤气厂公园改造经验的启发，这个项目听取了不同领域专家的意见，包括化学工程师理查德·布鲁克斯。此外，这个项目又一次提出运用生物修复技术治理地表污染（这个建议明显地违反了加州的法律，法律要求任何被污染的土地都必须隔离在一层泥土之下）。虽然只进行了简单的预测，但哈格的方法是帮助土地中天然存在的消化者充分发挥它们的作用。这个项目还扩展到了各种相互关联的层面，它实际上是形成了一个被自然本身激发的有机体。城市公园的典型功能和自然保护区是一致的，地形地貌、自然雨水、海水和植被在其中扮演了主导角色。哈格在设计中考虑了这些，使场地足够健康去支撑。项目结果形成了一个开发城市和工业废弃地潜力的生态机器，使路径、滨线和植被都作为项目的构成要素。当哈格在 0.4 公顷的场地实验他所提出的方法时，开拓性的方案再次被试验。

As in most American cities that give onto the ocean, the waterfront at Berkeley was not initially occupied by housing but by commercial and industrial activities which exploited the site's easy links with both the port and the hinterland beyond. Over time, this industrial development created a practically impenetrable belt of infrastructures that cut the city itself off from the sea. Sometimes, indeed, manufacturing activities extended even beyond the coastline, occupying artifical landfills created using domestic and industrial refuse.
It was 1987 when Richard Haag proposed that the vast 97-acre polder created alongside the Marina in this American university city should be redeveloped as a public park, transforming what had become an open-air rubbish tip into a new 'green lung.' Inspired by the experience acquired in creating the Seattle Gas Work Park, the project drew upon the collaboration of various experts, including the chemical engineer Richard Brooks. And, once again, it was proposed that ground pollution should be remedied *via* bio-remediation (a suggestion in clear opposition to Californian law, which requires that any polluted terrain must be insulated beneath a mantle of clay). Predicated upon simplicity, Haag's approach, however, was that the natural digestants (methanophilous bacteria) present in the ground should be assisted to perform their work. The project extended over various interrelated levels, in effect forming an organic whole inspired by Nature itself. Functions typical of a city park go together with those of a nature reserve, with leading roles being played by terrain contours, rain, sea water and vegetation; Haag includes the latter in the design wherever the ground is healthy enough to support it. The result is an ecological machine which 'exploits the potential' of urban and industrial waste, making it just as much a component of the project as pathways, shorelines and vegetation. The pioneering scheme is being re-examined as Haag proposed this approach be tested on a one-acre plot.

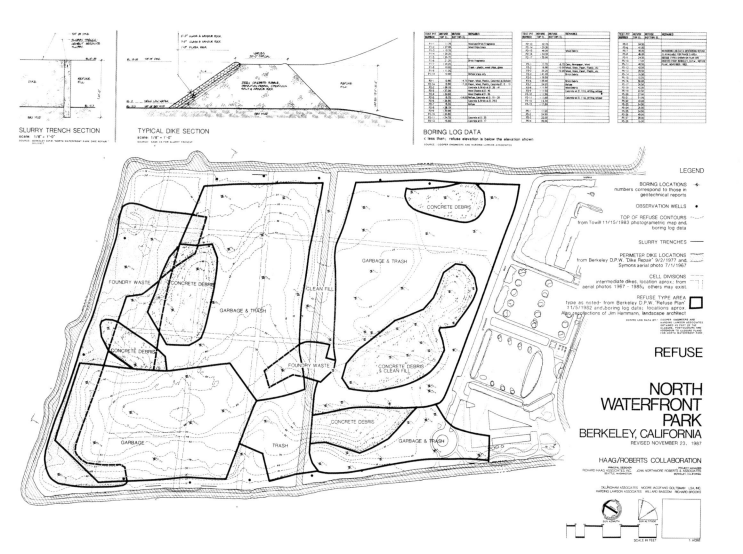

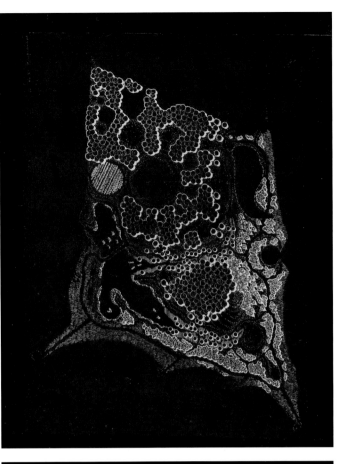
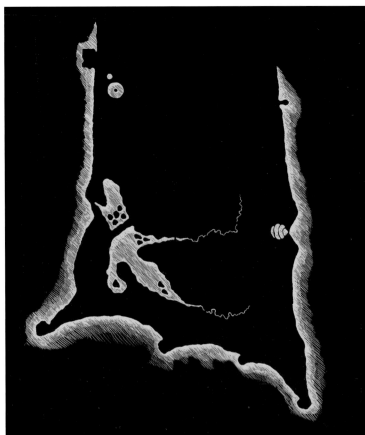
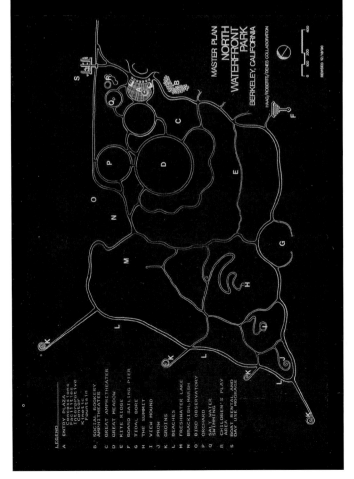
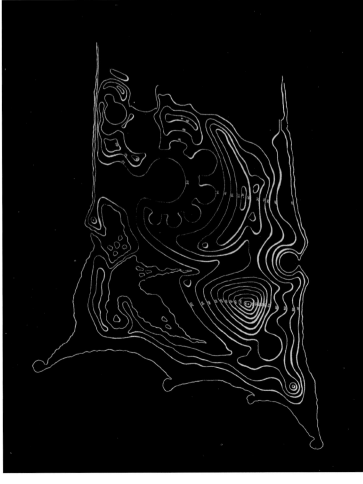

MASTER PLAN
NORTH WATERFRONT PARK
BERKELEY, CALIFORNIA

HAAG/ROBERTS/DELES COLLABORATION

REVISED 12/18/90

LEGEND
A ENTRY PLAZA
  Facilities
  Concessions
  Interpretive
  Exhibits
  Kinetic
  Fountain
B SOCIAL ROOKERY AMPHITHEATER
C GREAT AMPHITHEATER
D GREAT MEADOW
E KITE RIDGE
F BOARD SAILING PIER
G TIDAL BORE
H THE SUMMIT
I VIEW MOUND
J PROW
K GROINS
L BEACHES
M FRESHWATER LAKE
N BRACKISH MARSH
O BIRD OBSERVATORY
P ORCHARD
Q SALT WATER SWIMMING
R CHILDREN'S PLAY AREA
S BOAT RENTAL AND DAY USE MOORAGE

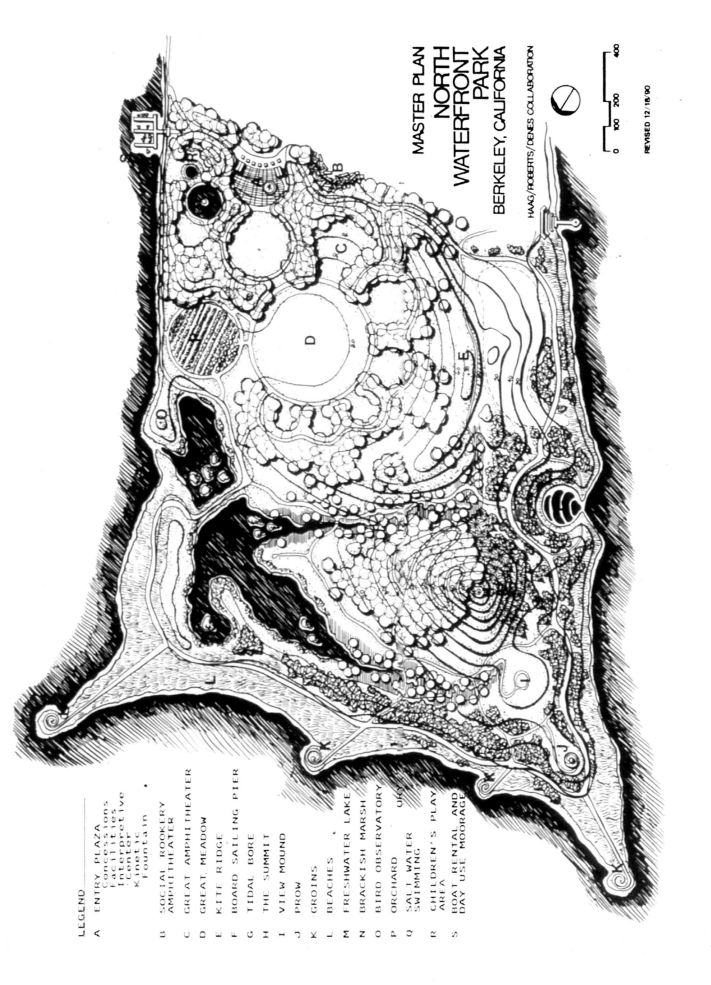

3.

案例研究之家

北纬：37°32′40.00″
西经：122°30′36.00″

地点：阿瑟顿，加利福尼亚州
时间：1957年
状况：未建成
项目设计：唐·克诺尔
项目经理：理查德·哈格
设计团队：理查德·哈格事务所
项目委托方：Entenza
工程设计：John Brown
制图方：Richard Haag
场地面积：0.40公顷（1英亩）

理查德·哈格本人发起了这个研究项目——当他在1957年草拟项目时，他还拥有加利福尼亚州伯克利的工作室——这对了解理查德·哈格的工作进展很重要。建筑本身是唐·克诺尔完成的，豪华的住宅反映了国际风格的经典模式。当时正值哈格被委托了一个艰巨的任务，就是使住宅的室内与室外空间的相互关系和谐地结合起来。在他的作品中，景观设计师从在日本的经历中获取构思方案的灵感。这样，建筑设计上的极简主义与透视和材料的使用相冲突，与当时在美国流行的审美观念大相径庭。房子被分成两个独立的部分，就好像被围绕它的白砂石从地面分离。昼间使用的空间渗透到外部景观，成为沉思的空间。孩子们可以在沙子上玩耍（从附近蒙塔拉的海滩，至旧金山以南都有这种沙子），沙子就铺在孩子们卧室正对的院子里。但是这些玩耍的地方也是沉思的地方，被定义为有点像禅园。布局几何对称的、被蔓延的草坪遮蔽了一半的游泳池通过一个桥到达，并由远处的竹帘提供挡风保护。四周分布着金丝桃属植物，以延伸和混合着贫乏的地形，不要形成一个明显的划分边界。在这个没有实施过的项目中，哈格渗入了自己对传统日本庭园个性化而有效的理解。不仅作为一种启发方式具有严密性和规律性，而且作为一种日常生活的准则具有秩序性和功能性。

Richard Haag himself indicated this study project – drawn up in 1957, when he still had his studio in Berkeley (California) – as essential to understanding the development of his work. The architecture itself was by Don Knorr, with this luxury house reflecting the classic schema of the International Style. Haag was commissioned with the precise task of orchestrating the interrelation between the internal and external spaces of the house. In his work, the landscape designer chose to drawn upon schema inspired by his experience of Japan. Thus the fake minimalism of the architecture clashes with a use of perspective and materials in sharp contrast with the aesthetic ideas then prevailing in America. Divided into two autonomous parts, the house seems to be separated from the ground by the outline of white gravel that surrounds it. The daytime spaces are permeable to the external landscape, becoming spaces of contemplation. Children can play in the sand (from the nearby beach of Montara, just to the south of San Francisco) which is laid out in the courtyards onto which their bedrooms open. But these play areas are also places for contemplation, defined like little Zen gardens. Half concealed by expanses of lawn laid out in geometrical symmetry, the swimming pool is reached by a bridge, whilst protection against the wind is provided by distant curtains of bamboo plants. All around, hypericum is laid out so as to form a border but not a sharp division, extending and blending with the arid terrain beyond. In this project, which was never implemented, Haag develops his own personal and efficacious interpretation of the historic Japanese garden. Not only is there rigour and discipline as a means to illumination, but also order and functionality as a basis for everyday living.

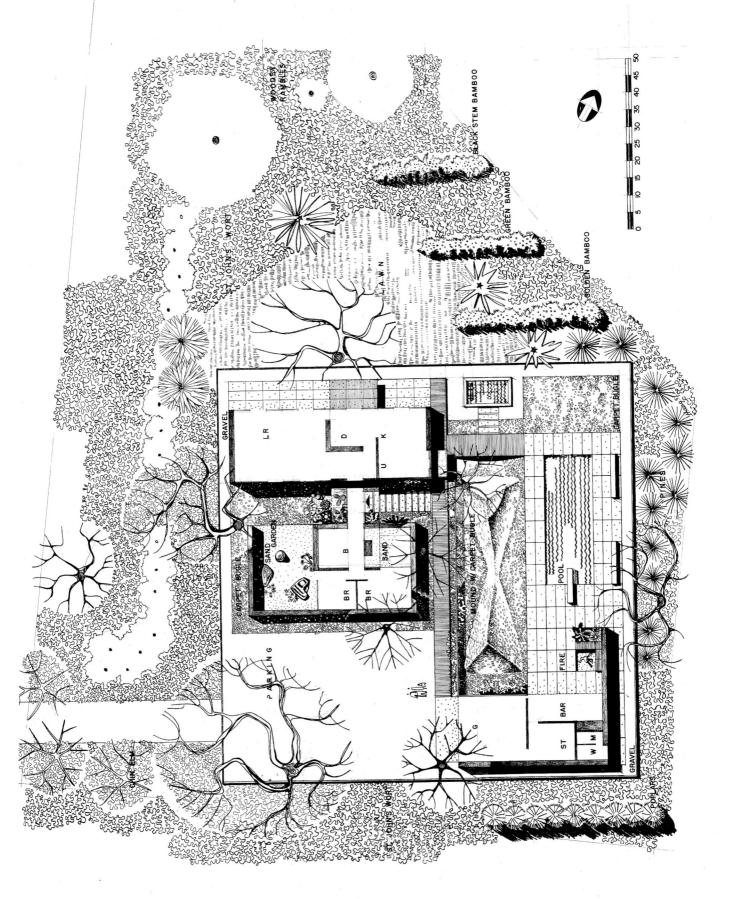

4.

杜克斯办公室

北纬：37°35′57.00″
西经：122°23′12.00″

地点：米尔布雷，加利福尼亚州
时间：1957～1960年
状况：已建成
项目设计：唐·克诺尔
项目经理：理查德·哈格
设计团队：理查德·哈格事务所
项目委托方：Dux Furniture Company
摄影师：courtesy RHA
场地面积：100平方米（4000平方英尺）

　　唐·克诺尔的建筑设计与理查德·哈格的景观设计又一次在米尔布雷结合在一起（又位于加利福尼亚半岛并终止于旧金山）。一个平顶方形建筑，即杜克斯家具公司的商业楼宇，被设计于1957～1960年，再次体现了建筑的理性趋势。例如，克诺尔用垂直排列的木材板装饰外墙，木材的红色调创造了一种时而温暖的效果，色彩的灰色阴影带来了一种时而更中性的效果。哈格把建筑当作背景理解，背景衬托他的景观概念更突出。仅仅通过包裹着建筑的表面展现出来的几个特点，就拓展了原本狭窄的空间。有时矿石（覆盖了一层白色的河卵石），有时植被（布局成一个紧密的草坪），地面变成数量有限的植被和树木的基础。附近的停车场种植着两排白桦树，伴随着红枫和绿色灌木对比性地布置在木质墙面前，留下斑驳的色彩。在一种超现实的绘画中，黑色的方形石头作为抽象的特征，分布在大片的鹅卵石中。它们闪亮的反射与灰白色的背景形成鲜明对比。他们似乎与浓厚丛生的绿草和贵族般纤细的木贼属植物建立了一种慎重的对话。在从办公室俯瞰的小院子里，常规以方形为基础的庭园布局被一个圆形石喷泉的存在打破。它的垂直喷水给予了整体一个恰当的深度，增加了布局进一步的尺度感。哈格又一次发展了日本的风格（重复、惊喜和数量有限的部件）去创造一个充满活力、非常现代的环境。几乎可以说是超越时代的环境，预示着三千年前的产物。黑色石头是被破译的信息。仅仅几年以后，斯坦利·库布里克呈现了令人不安的黑色单块巨石于他的2001年视觉杰作：《空间的奥德赛》（1968年）。也许这些方形的黑色石头可以被视为预示未来某种事物的降临。

Don Knorr's architecture was once more combined with Richard Haag's landscape designs at Millbrae (again on the Californian peninsula that ends in San Francisco). A flat-roofed square building, the commercial premises for Dux, a furniture company, were designed in the years 1957-1960 and again exemplify the rationalistic trends in architecture; Knorr, for example, faces the outside walls with timber slats organised to form a vertical pattern, creating an effect that is at times warm, thanks to the red tones of the wood, at times more neutral, thanks to greyish shades of colour. Haag interprets the building as a background, the backdrop against which his own notion of landscape will stand out. Just a few features are laid out across the surfaces that surround the building, serving to expand a space which is itself quite narrow. At times mineral (covered with a layer of white river pebbles), at times vegetable (laid out as a close-cut lawn), the ground is the base from which emerge a limited number of plants and trees. Near the car parks are double rows of birch trees, with bursts of colour from the red maples and a few green shrubs set in contrast against the wooden walls. As in a sort of Suprematist painting, square black stones are laid out as abstract features within the expanses of pebbles. Their shiny reflections are in sharp contrast with the grey-white background, and they seem to establish an discreet dialogue with the green of the thick tufts of grass and the aristocratic slenderness of the equisetum. In the small courtyard overlooked by the offices, the regular square-based layout of the garden is broken by the presence of a circular stone fountain. Its vertical jet of water gives the right degree of depth to the whole, adding a further dimension to the layout. Once again, Haag develops upon Japanese stylemes (repetition, surprise and a restricted number of components) to create a dynamic and very contemporary environment. One might almost say this is an environment ahead of its time, heralding those of the third millennium. The black stones are messages to be deciphered. Just a few years later Stanley Kubrick would present us with the disturbing presence of the black monolith in his visual masterpiece *2001: A Space Odyssey* (1968). Perhaps these square black forms might be seen as a foretaste of things to come.

5.

罗斯福纪念碑竞赛

北纬：38°53′2.18″
西经：77°2′38.54″

地点：华盛顿特区
时间：1958 ~ 1959 年
状况：未建成
项目设计：亚伯拉罕·W·盖勒（1959）
项目经理：理查德·哈格
设计团队：理查德·哈格事务所（劳瑞·奥林，弗兰克·洛克德）
制图方：Richard Haag, Frank Lockfeld
建模：IMPEI's model shop
艺术家：Laurie Olin
摄影师：courtesy RHA
场地面积：30 公顷（75 英亩）
建造费用：800 万美元

在华盛顿特区的富兰克林·德拉诺·罗斯福纪念碑设计竞赛见证了理查德·哈格搬到西雅图后的首次成功。1958 年，他的竞赛作品在波托马克河与潮汐盆地之间的一片狭长用地上，将建筑与景观设计结合在一起。整体上可被视为一个充满动感的竞赛作品。它将有点轻微起伏的场地引导至纪念性建筑本身。树木选择了开花结果的树木（如樱桃和苹果），从而创造出小树林繁花盛开的灿烂景象，效仿在日本景观设计中常见的活力。广场的建筑是个黑色花岗石的巨大基座。基座的顶上是由 51 根柱子组成的优雅结构，从中间的高度绑在了一起，使人联想到土著印第安人帐篷的形态。在距离政府中心不远的地方，哈格创造了一个作为美国情感象征的景观，直插云霄的高杆由相互间的联合而彼此支撑，清晰地隐喻着合众国。然而，纪念碑不会被兴建。一年以后，盖勒加入到第二阶段的竞赛（1959 年），将哈格作为景观建筑师。结果（见对面页）比起第一阶段，减少了建筑的冲击力。以紧凑的砌块为突出特征的第一阶段设计，在这里变成了复杂的几何结构，利用周边环境制定了规则。比起先前的设计少了运动中的协调感，哈格的设计旨在提供一个充满活力的建筑布局，其灵感来源于具有 20 世纪 70 年代特色的塑性建筑形式。

The competition for the Franklin Delano Roosevelt Monument in Washington D.C. saw Richard Haag's first success after his move to Seattle. Presented in 1958, his competition proposal combined architecture and landscape design within a long tongue of land between the river Potomac and the tidal basin. The whole is envisaged as a dynamic composition, with slight depressions and rises in the site leading up to the monumental building itself. The trees are flowering fruit-trees (cherry and apple), thus creating small woods whose brilliant springtime blossom echoes the dynamism to be seen in Japanese landscape design. The square building is also a massive pedestal in black granite surmounted by an elegant structure of 51 poles that are bounded together at mid-height in a way that recalls the form of a Native American tepee. So, just a short distance from the centre of government, Haag created a landscape that is an emotive symbol of America itself; the tall poles reaching into the sky are supported by their very union with each other – a clear metaphor for the United Nations. However, the memorial would not be built, and a year later Geller would take part in a second – phased – competition (1959), counting upon Haag as landscape architect. The result (see opposite page) is less architecturally striking than the first. The compact block that was the characteristic feature of the first design here becomes a complex geometrical construct that imposes the rules for the surrounding area. Less harmonic in movement than his previous solution, the design that Haag proposes here aims to provide energetic emphasis for an architectural composition inspired by the interest in plastic form which was so characteristic of the '70s.

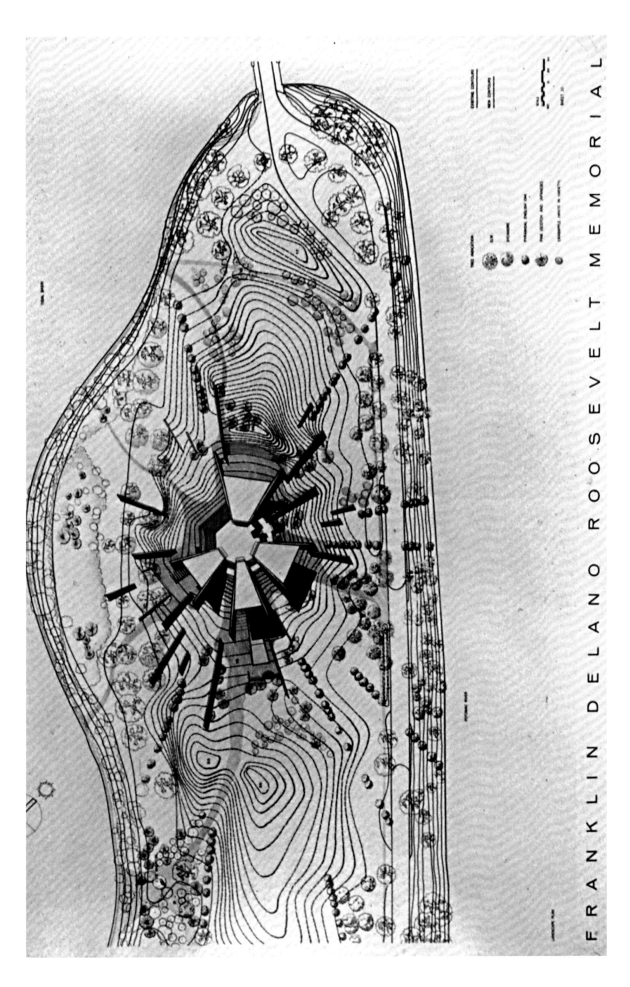

FRANKLIN DELANO ROOSEVELT MEMORIAL

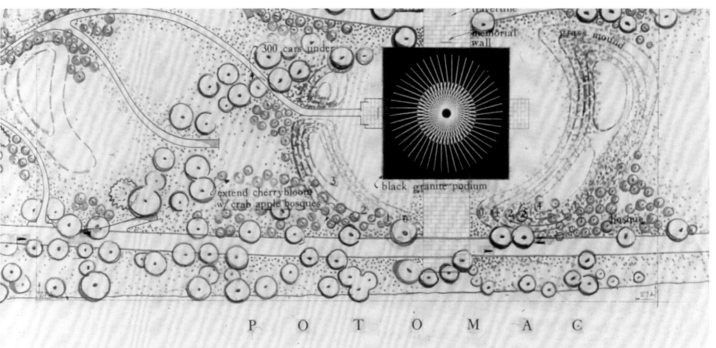

ROOSEVELT MEMORIAL

6. 弗莱艺术博物馆

北纬：38°53′2.18″
西经：77°2′38.54″

地点：西雅图，华盛顿州，特里大街704号
时间：1994～1997年
状况：已建成
项目设计：里克·桑德伯格、奥尔森·桑德伯格
项目经理：理查德·哈格
设计团队：理查德·哈格事务所
项目委托方：Frye Art Museum
主要承建方：Jolly Miller Concrete
制图方：RHA
摄影师：courtesy RHA, Luca M.F. Fabris

1994年，西雅图最重要的私人博物馆委托哈格设计入口和庭园的布局，为藏品的进一步增加提供必要保证。这位景观建筑师充分利用绿化，用一片树木帷幕构筑博物馆入口（博物馆对面是一个使用功能良好但毫无辨识度的停车场）。这样也能够装饰道路，提供一个整体上让人熟知的欧洲城市风貌。哈格还运用水元素，设计了一个反射阳光的水池，赋予了朴素庄重的建筑新的亮点。他采用的第三种设计元素——混凝土，使他能够真正达到独特的境地：在小院落里，从咖啡厅和博物馆的一些房间都可以看到的角度，他布置了一个粗糙混凝土制的巨大球体来象征世界的不稳定。在哈格访问日本40年后，他重新整理了在那里学到的设计规则。暴露在日晒雨淋，严寒酷暑下，这个超乎一般尺度的物体放大了小小的开放空间的尺度。这个不知从何处坠落的球体与上天之间进行着一场对话，不断侵入其表面的青苔强调了景观的诗意。在它的下面，大片百合花草地作为基础，继而扩展成不连续的铺地空间。哈格运用喷射混凝土制作这个巨大球体，利用球体不平整的表面种植一种石生藓类植物：一种随季节改变颜色的苔藓，在这里生存必须没有任何形式的人工灌溉。植被和石头的关系随着人的视点不同而改变，揭示着不断转变的叙事关系。从院子内部，人们要看到这个与宇宙自由对话的球体，必须透过铁丝网，面朝着另一个世界——人类的世界。这是一个概念艺术的真实榜样，也是理查德·哈格最杰出的作品之一。

In 1994 the most important private museum in Seattle commissioned Haag to design the layout for the entrance and courtyard of new premises made necessary by further additions to the collection. The landscape architect made use of greenery, marking out the entrance (opposite the museum is a useful but undistinguished car park area) by a curtain of trees; this also serves to adorn the road itself, giving the whole a European urban appearance that has a familiar feel to it. Haag also used water, designing a pool whose reflected sunlight would cast new highlights across the sober architecture of the building. The third element he chose to use – concrete – enabled him to come up with something truly unique: in the small courtyard, accessible from the cafeteria and visible from some of the museum rooms, he placed an enormous sphere of rough-cast concrete, which stands as a symbol of the precariousness of the world. Forty years after his visit to Japan, Haag was reinventing the rules learnt there. Standing exposed to sun and rain, heat and cold, this outsize feature serves to amplify the scale of the small open-air space. There is an on-going dialogue between this sphere, which has dropped to earth from who knows where, and the sky above. The poetry of the object is underlined by the moss that indefatigably invades its surface. Beneath it an expanse of lily turf serves as a base and then expands into the non-continuous space of the paving itself. Haag made the sphere out of shotcrete, taking advantage of the uneven surface to plant *ceratodon purpureus*, a moss that changes colour with the seasons and here must survive without any form of artificial irrigation. The relation between the vegetable and the mineral changes with one's point of view, revealing a continually shifting narrative. From within the court, one sees how the sphere, which is in free dialogue with the cosmos, must look out towards the world beyond, the world of humankind, through a network of iron railings. This is a true example of conceptual art – and one of Richard Haag's very finest design projects.

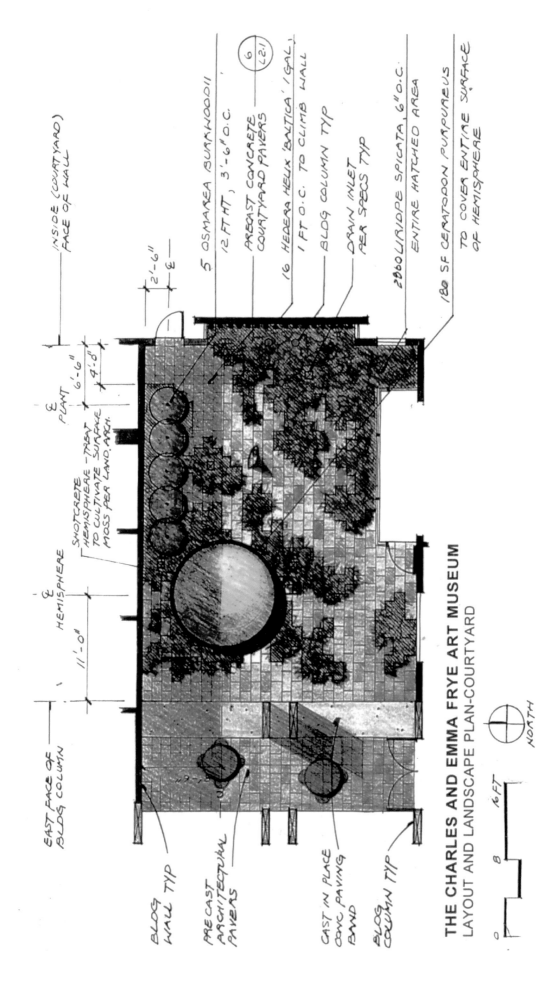

THE CHARLES AND EMMA FRYE ART MUSEUM
LAYOUT AND LANDSCAPE PLAN-COURTYARD

Richard Haag Associates, Inc.

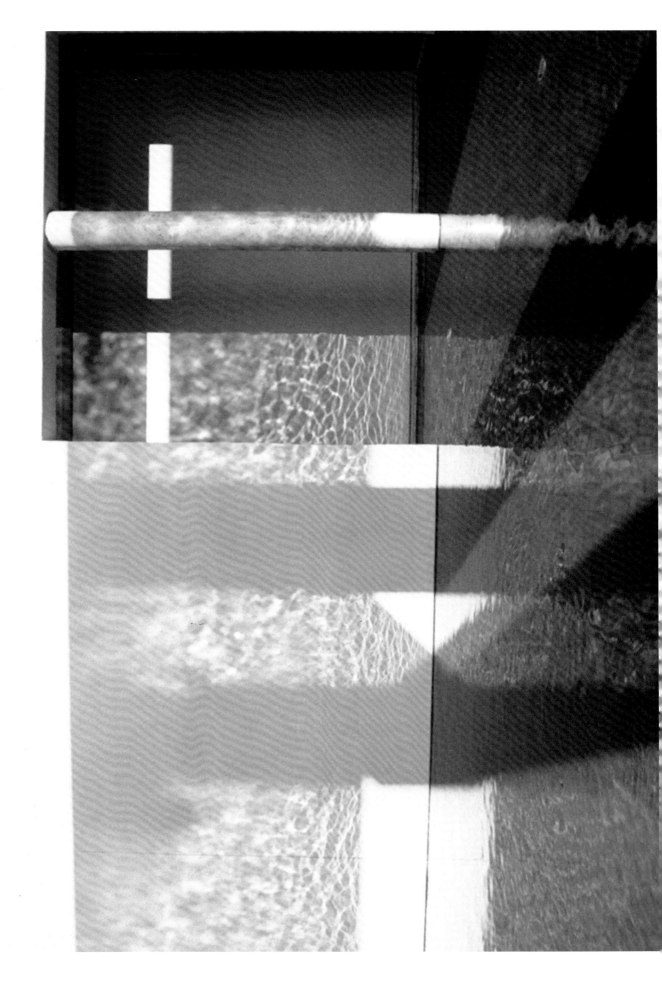

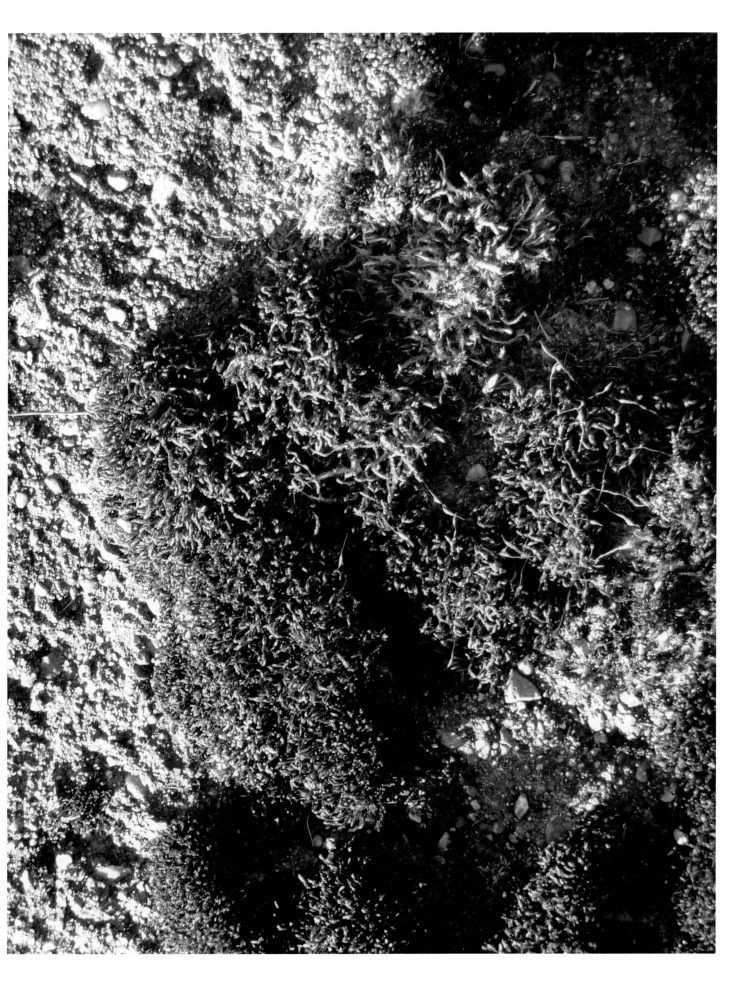

7.

巴特尔纪念馆

北纬：47°39′35.78″
西经：122°17′5.24″

地点：西雅图，华盛顿州，第41东北大街4000号
时间：1965～1971年
状况：已建成
项目设计：雷拉莫尔·布雷恩，布拉迪·约翰森，戴维·赫德梅克（建筑设计）
项目经理：理查德·哈格
设计团队：理查德·哈格事务所（罗伯特·汉娜，第一阶段；克雷格·坎贝尔，第二阶段）
项目委托方：Battelle Memorial Institute
摄影师：courtesy RHA, Luca M.F. Fabris
场地面积：7公顷（18英亩）

1960年代后期对公众开放的巴特尔纪念馆，现在作为塔拉里斯会议中心而闻名，是西雅图第一个这样的会议设施。它是由一个演讲大厅，自修室和临时住所组成的综合体，其中的临时住所是为附近的华盛顿大学的研究者和学者们提供的。哈格利用场地目前作为水草地的斜面，排掉多余的水，形成一个有着曲折轮廓的巨大人工湖。整个布局取决于这个水池，它提供了被建筑整体界定的视线的透视消失点。在湖的更远处，一片树林修复了这个广阔的大学综合体（总计18英亩）的天然环境。在1992年的设计师评估中，哈格的布局仍然保持着强有力的效果，将复杂的校园作为一个整体来理解，它提供了一个视觉中心。在入口处，一个人进入了树林后来到高处，会最终发现整个校园的布局。就像在哈格所有的大型项目中一样，不同的高差和坡度创造了整体感受。在各种各样的建筑之中，一系列的庭园空间为人们提供了令人愉悦而宁静的领域，这些铺地的表面巧妙地使用植物来强调。建筑物和开放空间在水平线上依次展开，创造了看似相互流动的环境。哈格采用的构图语言表现在细节的重要性上，这大部分归功于使用像混凝土这样的日常材料。这里显示日本设计本身的并不多，而是经过卡洛·斯卡帕解释的日本设计。细节变成存在的理由、手段以及各个方面。草地的边缘，长长的台阶，相互交错的石头和绿化，这些都是这个项目的精髓所在。此外，桥和岸边的大垂柳与湖一起俨然形成一幅画卷，而哈格已经设法把它变为现实了。

Open to the public in the late 60s, the Battelle Memorial Institute – now known as the Talaris Conference Center – was the first such conference facility in Seattle; it comprises a complex of lecture halls, study rooms and temporary accommodation for researchers and scholars associated with the nearby University of Washington. Haag makes use of the dip in the terrain, an actual water meadow, to drain off the excess water and form a large artificial lake of sinuous outline. The whole layout hinges upon this water basin, which provides a perspective vanishing-point for sightlines defined by the architectural ensemble. Beyond the lake, a wood restores the natural setting of this vast university complex (totalling 18 acres). Reviewed by the designer in 1992, Haag's layout is still powerfully effective, providing a visual key for an understanding of the complex as a whole. At the point of access, one enters into the wood, rising upwards to finally discover the entire layout of the campus. As in all Haag's large-scale projects, are these differences in level and gradient those serve to create a sense of the whole. Amongst the various buildings, a series of courtyard spaces provide areas of convivial calm, where the paved surfaces are underlined by the skilful use of vegetation and plants. The buildings and open spaces follow on from each other in horizontal lines, creating environments that seem to flow into each other. The compositional language adopted by Haag plays upon the importance of details, which here – largely thanks to the use of such everyday materials as concrete – suggests not so much Japanese design itself but rather Japanese design as interpreted by Carlo Scarpa. Details become the *raison d'être*, the measure, of all things. The edges of the lawns, the long steps, the intercutting of stone and greenery – these are the very essence of the entire project. Furthermore, with its bridge and large weeping willows, the lake stands as a stock image which Haag has managed to transform into a reality.

Battelle
Seattle Research Center

MASTERPLAN UPDATE 1992

8.

美洲园艺博览会竞赛

北纬：39°57′57.91″
西经：82°57′18.12″

地点：哥伦布市，俄亥俄州
时间：1992 年
状况：未建成
项目经理：理查德·哈格
设计团队：理查德·哈格事务所
项目委托方：Ameriflora 1992, Inc
艺术家：Kent Dickson, John Koepke
场地面积：2 公顷（5 英亩）

在富兰克林举办的园艺博览会是哥伦布市（俄亥俄州）作为纪念哥伦布发现美洲500周年系列活动的一部分。哈格为此而创作的设计图展现了他在绘图上的所有技巧。事实上，这位设计师始终随身携带着铅笔和素描本。有时用铅笔、有时用彩色蜡笔，他通过充满冲动和激情，并且设计意图非常清晰的草图来表达他的灵感与想法。哈格也喜欢用铅笔草草记下自己的注释，在他干净手稿上出现的厚重线条，往往是物质上表达强调的转变。这些都可以反映在旁边的草图页，在这个赢得竞赛的设计作品中，构思、透视图、造型相互穿插，进而在图稿上体现出来。这些草图和画稿为我们对设计者启发性的工作方式提供了精确的视点，也让我们能够想象那些最终成型的设计理念诞生的独特瞬间。这些精心组织地配合着概括性图解的成果展示，是对那些含糊不清的设计过程的独家揭秘。当你看到这些草图时，你会体会到一个离经叛道的设计者的工作方式。这是一个用自己作品说话的景观艺术家，他的工作就是体现自然。比如，在1992年美洲园艺博览会的竞赛方案中，他利用光与影、水与风的相互作用，创造了一系列的树林与花园。在植物迷宫和巨大的玫瑰花园之间，他设置了一个冥想空间，供人们反省自己。在遥远的尽端，高大的树木形成了"自然神殿"，通过绿化构成的"风廊"到达。这些空间不是用来庆祝征服的，而是高贵地阐明在这里分享的发现。

Created for a horticultural exhibition held in Franklin, Columbus (Ohio) as part of the 500th anniversary of Columbus' discovery of America, Haag's sketches reveal all his skill as a draughtsman. In effect, the designer keeps a pencil and pad along with him at all times. Sometimes using lead pencil, sometimes coloured crayon, he expresses his ideas in sketches that are impulsive and full of passion but also extremely clear in intent. Haag also uses pencil to jot down his notes, the very thickness of line in his neat script giving material expression to shifts in emphasis. All of this can be seen in this page of drawings (alongside), in which ideas, perspective views and forms develop from each other to then materialise in the winning design submitted to the competition. These sketches and drawings provide a precious insight into the way the designer works heuristically towards that unique moment in which the final idea takes form – an intimate 'breakthrough' which is only vaguely reflected in the carefully-ordered (and necessarily schematic) depictions to be seen in the final design. When one looks at these sketches one gets some notion of the working methods of a designer who is an iconoclast *malgré lui*. This is a landscape artist who speaks through his own works, his own work upon Nature. So, for example, at Ameriflora '92 he created a sequence of woods and gardens which exploit the interplay of light and shade, of water and wind. Between the plant labyrinth and the large rose garden, he sets a space of meditation, in which the created world reflects/reflects upon itself. At the far end, a crown of tall trees form a Temple of Nature, reached *via* a 'wind gallery' of greenery. These are not spaces intended to celebrate a conquest; they are the dignified interpretation of a discovery that is there to be shared.

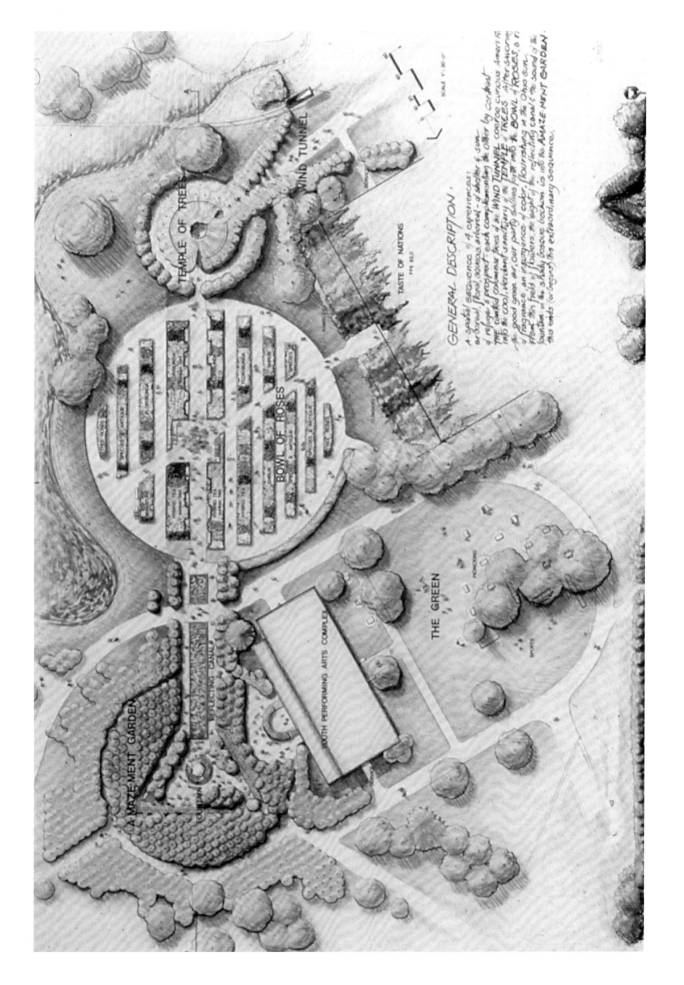

9.

乔丹公园／埃弗里特码头公园

北纬：48°0′13.50″
西经：122°13′18.00″

地点：埃弗里特市，华盛顿州
时间：1972 年
状况：已建成，并于 2005 年被拆除
项目经理：理查德·哈格
设计团队：理查德·哈格事务所（戴勒·丹尼斯）
项目委托方：Port of Everett
摄影师：Mary Randlett, courtesy RHA
场地面积：0.33 公顷（0.84 英亩）

埃弗里特市是一个位于西雅图北边俯瞰普吉特海峡的小镇。在埃弗里特码头附近，乔丹公园好多年前被推土机推平，已经不存在了。它是理解哈格独创的景观设计方法的集中代表作，可惜它沦为码头和新基础设施发展（通常使用的说法）需要的牺牲品。追溯到 1970 年至 1972 年期间，这个项目非常重要，因为它和同期的煤气厂公园（某种程度而言是相互补充的），以及另外两个无可争议的杰作：罗伯特·史密斯森在大盐湖的螺旋形防波堤（犹他州，1970 年）和罗伯特·莫里斯在艾居默伊登的瞭望台（荷兰，1971 年），开辟了将景观建筑转变为大地艺术的途径。在这四个案例中，由砂石组成的大地本身扮演了领导的角色，被塑造而成雕塑。这种谦卑而重要的物质，启动了一种新的艺术形式。

从尤卡坦参观回来后，理查德·哈格在埃弗里特的设计中，运用了非常刚性的构造形式，名副其实的截头棱锥体。这些不同高度的平台，被用来作为观景瞭望台。景观建筑师使用了大约 3800 多立方米（5000 立方码）从码头挖出来的砂石来完成这些构造形式。就像在煤气厂公园采用的方式一样，项目将碎石和废弃物变成了景观和建筑。哈格将覆有短草的金字塔平台分散式地布置，以便创造给家庭使用的适中空间——比如，野餐草地——或者人们在草地上或坐或躺的其他场所，以给予人们时间去再次发现大量现代世界中被忽视的景象（黎明、黄昏、火山的远影）。成排的白杨木用作构图的幕帘，就像是这构图的帘幕，开合之间，步移景异。公园虽小，却个性十足。就像一本打开的书，它能够孕育强烈的感情和反应。不幸的是，我们的世界目光太短浅而不能保留这个重要的公共财富。

Located near the Marina of Everett – a small town to the north of Seattle overlooking Puget Sound – Jordan Park no longer exists as it was bulldozed a few years ago; a work that is central to an understanding of Haag's original approach to landscape design, it fell victim to the need to 'develop' (that term is always used) the marina and its new infrastructures. Dating from 1970-1972, this project was important because contemporary with the Gas Works Park (to which in some ways it was complementary) and to two other indisputable masterpieces that opened the way for the transformation of landscape architecture into land art: Robert Smithson's Spiral Jetty at Great Salt Lake (Utah, 1970) and Robert Morris's Observatory at Ijumuiden (Netherlands, 1971). In all four cases, the earth itself – made up of sand and stone – played a leading role, being modelled to become sculpture; humble, yet essential, materials generated a new form of art.

Having just returned from a visit to the Yucatan, Richard Haag used very rigid tectonic forms in his designs for Everett; veritable truncated pyramids, these would – thanks to their differences in height – serve as observation platforms. The landscape architect used around 5,000 cubic yards of materials dredged from the marina itself to construct these forms; as happened at Gas Works Park, the project transformed rubble and waste into landscape and architecture. Covered with closely-mown grass, these pyramids were distributed by Haag so as to create intermediate spaces for family use – for example, the picnic lawn – or other areas where one might sit or lie out on the grass to allow oneself time to rediscover a whole number of things which the contemporary world fails to notice (dawn, sunset, the distant view of volcanic mountains). Rows of poplars served as a curtain for the composition, opening and closing to frame the views beyond. Though small, this park had a very clear identity; an open book, it could nurture powerful emotions and reactions. Unfortunately our world proved too short-sighted to preserve this important public treasure.

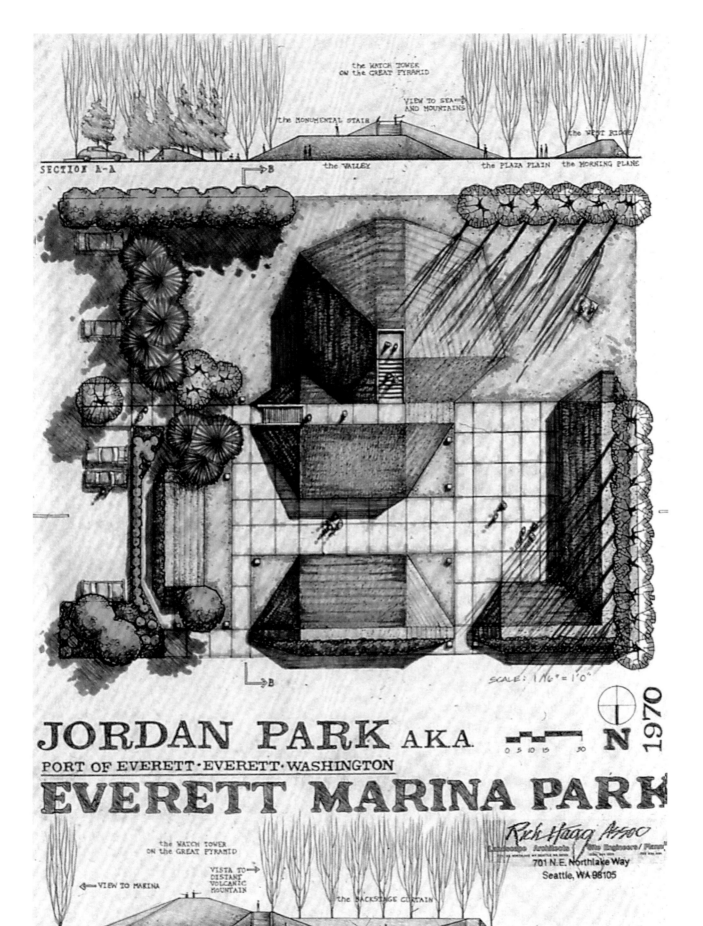

10.

加利福尼亚大学伯克利分校林间纪念广场

北纬：37°52′23.00″
西经：122°15′33.90″

地点：伯克利，加利福尼亚州，大学车道
时间：1994～1996 年
状况：已建成
项目经理：理查德·哈格
设计团队：理查德·哈格事务所
项目委托方：University of California, Berkeley
场地面积：2 公顷（5 英亩）
建造费用：70 万美元

为了纪念在第二次世界大战中做出贡献的教员、学生和职工，建成的开放空间位于校园中心，利用大学车道形成的巨大弧形，正对着 Doe 纪念图书馆的宏伟建筑。理查德·哈格设计了新的纪念性景观，以一片开阔的、轻微到几乎觉察不出的斜坡，将视线引向一排遮挡着后面道路的高大树木。纪念性的林间空地因而形成了包容空间，像真正的房间一样，可以一人独享，也可以众人同乐。图书馆正对的大型开放区域，也是一个可容纳 3500 人的圆形剧场。在萨瑟路街角，一座小型的圆形喷泉隐藏在一圈树阵中，构筑了一个平静而安宁的小空间。

实际上，项目在实施中有点偏离哈格最初的方案。特别是在这个巨大的开放广场中间和周边种植树木的数量，存在实质性的减少。

In remembrance of the faculty, students and staff who served in the Second World War, the open space within the heart of the university campus exploits the large bend formed by University Drive right opposite the imposing building which houses the Doe Memorial Library. Richard Haag designed the new commemorative landscape with a broad sweep, the gentle, almost imperceptible, slope leading down to a curtain of tall trees that conceal the road beyond. The 'memorial glade' thus formed contains spaces that are veritable 'rooms' where one can rest alone or in company. As for the large open expanse opposite the Library, this is also an amphitheatre which can hold up to 3,500 people; at the Sather Road corner is a small circular fountain hidden away within a circle of trees, offering a small space of peace and tranquillity.

The project as actually carried out involved some slight variations to Haag's original plans; in particular, there was a substantial reduction in the number of trees planted in/around the large open glade.

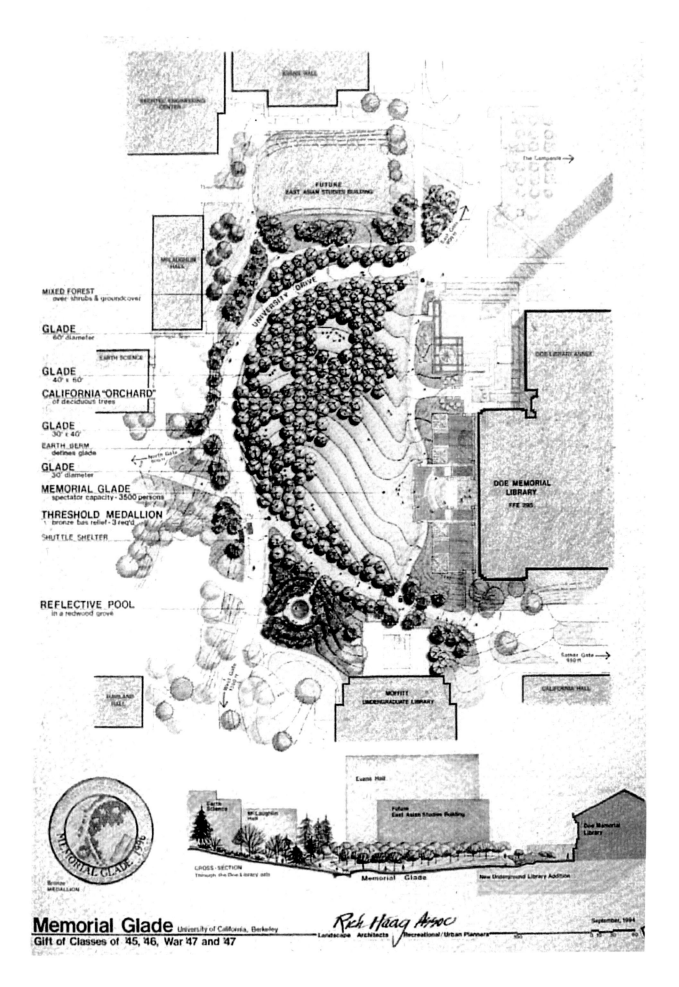

布洛德尔保护地

北纬：47°42′23.09″
西经：122°32′44.80″

地点：布兰布里奇岛，华盛顿州，东北道尔芬大道 7571 号
时间：1969 ~ 1986 年
状况：已建成
项目经理：理查德·哈格
设计团队：理查德·哈格事务所（马伽瑞特·马科斯维尔，约翰·本奈特）
项目委托方：Prentice Bloedel; Dick Brown (direttore Bloedel Reserve director)
摄影师：Mary Randlett, Luca M.F. Fabris, RHA
场地面积：57 公顷（140 英亩）

布洛德尔保护地项目就是一个关于委托方和设计师之间的友谊和理解的故事。事实上，理查德·哈格与布洛德尔保护地真正的管理者普兰提克·布洛德尔保持着对话。普兰提克·布洛德尔是一个木材大亨，也是普吉特海峡最重要的家族之一的成员。他委托设计师在这片距离西雅图大约 45 分钟渡轮到达、位于布兰布里奇岛上的私人领地工作。这里原本就有公园和住区。设计师所需要做的，就是通过充满耐心的工作去复苏因为矫饰的设计而失去活力的景观。为了恢复破碎的自然肌理，他必须赋予这块土地规划新的理解和含义。

大约在 1969 年，哈格开始为布洛德尔工作。合作持续了近 20 年，直到业主去世（那时其他的利益方接管了这块地产，随后产生的决定值得高度质疑）。哈格在这里的工作，可以看作是对明显脆弱且不可持续的肌理，一系列强有力的决定性干预。事实上，多年以后几乎很难找寻到，这些在专业细节上干预的痕迹：现在的一切都显得那么自然，即使本来并不是这样。这些都是人工干预的结果，而不是天生的自然。

57 公顷土地的大部分，都是经过精心考虑而留下来的树林和牧草地。广大宽阔而带有轻微坡度的草地界定着地平线。哈格利用土地轻微的隆起，隐藏了抬高的别墅视线，拉伸了远景和距离。实际上，在整个园区内，你所能见的和所不能见的都是精心安排的剧目。这个预先构思好的方案中，哈格再次以他个性化的处理手法解决两个难题，创造了解决方案的新类别。他设计出了一系列的庭园，正是这让布洛德尔保护地名声大振。

第一个空间就是"平面花园"。这里哈格的任务就是隐藏位于用作客人住宿的木屋附近的游泳池。这个木屋是保罗·海登·柯克的作品，呼应原住民长屋的设计。景观建筑师选用两个正交的金字塔将泳池覆盖起来，一个顶端朝向天空，一个深入地里，就像镜子中彼此的

The whole story of the Bloedel Reserve project hinges upon the friendship and understanding between the client and the designer. In fact, Richard Haag talks with real admiration of Prentice Bloedel, a timber magnate and member of one of the most important families in Puget Sound, who commissioned the designer to work on this private estate on the island of Bainbridge, a 45-minute ferry ride from Seattle. The park and the residence existed already. What the designer had to do was work with patient care to reconstruct a landscape that had been deadened by mannered design; restoring the torn fabric of Nature, he had to give new sense and significance to the lay of the land.

Haag began working with Bloedel around 1969 and the collaboration would continue for almost 20 years, ending only with the death of the property owner (when other interests took over the estate, resulting in decisions that are highly questionable). Haag's work here can be seen as a series of powerfully decisive interventions upon an apparently weak and insubstantial fabric. In fact, years later it is difficult to identify all of these interventions in specific detail: everything now seems natural, even if it isn't. What we have here is the result of human intervention standing in the stead of Nature.

A large part of the estate's 140 acres was deliberately left to woodland and pastureland. Large expanses of gently sloping meadow serve to mark the horizon. This gentle swell in the land allowed Haag to conceal the view of the drive up to the villa and to extend vistas and distances. Indeed, throughout the estate there is a play upon what you see and what you don't see. Within this pre-established schema, Haag resolves in his own individual way two difficult problems and invents new types of solutions, creating that sequence of gardens for which the Bloedel Reserve is famous.

The first space was the 'Garden of Planes'. Here, Haag's task was to conceal the swimming-pool located near the wooden guest-quarters; the work of Paul Hayden Kirk, these were to a design that echoes that of the First Nations longhouses. The landscape architect chose to cover the pool with a pair of two orthogonal pyramids that stand as if mirror images of each other reflected in the surface of the terrain: the ridge of one points heavenwards, the tip of the other sinks into the netherworld. Made of granulated granite defined by a steel strip, this geometrical structure of clean efficacious lines expresses the balance which ex-

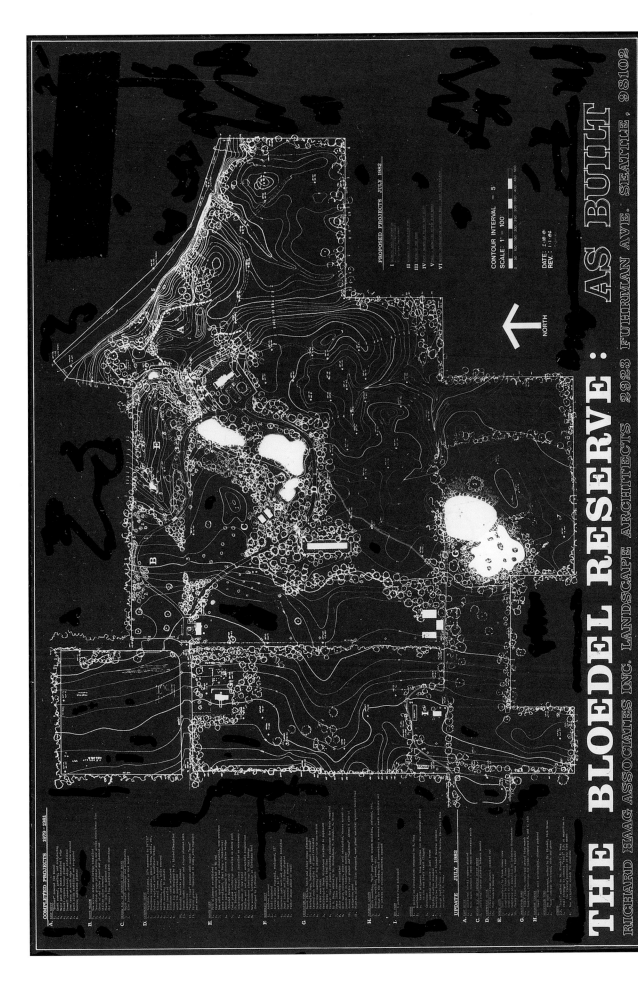

影像反射在地表上。由钢条限定的粒状花岗石构成，这个线条清晰有力的几何体表达了支配世界的不可见的力量之间的平衡。这里，哈格将他在京都银阁寺枯山水感受到的禅宗主题转化为西方的格调，在其中数学多于玄学。在周边，景观建筑师用草地布局了一种棋盘（来源于日本艺术的另一图形），创造出安静和谐的氛围。这一特征既与环抱在外的混凝土铺地相融合，也结束于长满蓝绿色牛尾草、布置着令人惊讶的奇怪石头的小山丘面前。哈格用存在的元素——混凝土铺地、红吠松，不知道什么时候留在这儿的巨大岩石——来组织空间，赋予空间新的活力。原先那些难以处理的、过分突出的东西，如今都成了单纯而庄严的抽象艺术作品。

再往前一点，你就置身于"创造自然的自然"，一个原始的苔藓园。这里仿佛一个原始森林，光线透过密林，投射出绿中带黄的颜色。人的脚步踏入这厚厚的苔藓几乎听不到声音。沿着小径再往前走，你会发现像被人遗忘似的倒下的树木，数百年的老树干看起来像一些古代寺庙的遗迹。哈格把这个空间描绘为"史前场所"。它是个迷人的地方，在这里人可以重新发现时间的感觉和它无情地流逝。而实际上，这些树干是1880年这里的森林采伐留下的。大约2200盘爱尔兰苔藓（珍珠草）被移栽到这里，森林因为移植在有些地方变得稀薄，在有些地方变得厚实，这些都不重要。实际上，这个地方简直是不可思议的神奇。

再往前就是"倒影园"，旨在创造一种位移的感觉；自然的无序现在要让位于均衡的有序。通过一个切入10英尺高红豆杉树篱的小开口，你可以看到一片天空，倒映在一个盛满深色水体的长长的（大约200英尺）水池中。无论是人们到这里沉思自省，还是享受光线的倒影，都不会否认布局上的理性。光线突然变得不可抗拒；周遭事物的常规尺度又强加给人绝对的命令，"清空你的大脑，不要犹豫，不要怀疑！"然而，即使在这里，哈格也只是在已经存在的事物上进行创作。他第一次来看

ists between the invisible forces that govern the world. Here, a Zen theme – which Haag had encountered at the Sand Garden of the Ginkakuji Temple (Kyoto) – is transposed into a Western key, wherein mathematics prevail over metaphysics. Around this, the landscape architect establishes an air of quiet harmony, using grass to lay out a sort of chessboard (another motif taken from Japanese art). That feature either blends in with the concrete paving which embraces it or comes to a halt before by a raised cone of land covered with blue-green fescue grass, from amidst which arises the surprising presence of an erratic boulder. Haag uses existing components – concrete paving, a red-barked pine, an enormous boulder left there who knows when – to organise the space and give it new energy. What might have been cumbersome and over-emphatic actually results in pure and sublime abstraction.

A few metres further on and one is in the midst of *Natura naturans*, in a proto-garden of moss. Filtered by what seems a primeval forest, the light here becomes greenish, almost yellow. Sounds are muted as one's footsteps sink into the deep moss. And as one proceeds along the path, one finds felled timber forgotten where it lies, trunks of centuries-old trees that look like the remains of some ancient temple. Haag describes this space as 'The Anteroom'. It is an enchanted place where one can rediscover a sense of time and its inexorable advance; the tree trunks, in fact, are those left over from the deforestation carried out here in 1880. It is of little importance that the moss was transplanted here (a total of 2,200 trays of Irish moss, *sagina sublata*) or that the forest has been thinned out at some points and replanted to make it more dense at others. This place is simply magical in effect.

A little further on is the Reflection Garden, which aims to create a sense of displacement; natural disorder now gives way to a perfect order of proportions. Through a small opening cut in a 10-foot-high yew hedge, one gains access to a portion of sky as reflected in a long (around 200-foot) pond full of dark water. Whether one comes here to reflect or to enjoy the reflections of light, the rationality of the layout cannot be denied. The light suddenly becomes irresistible; the regular scale of things imposes a categorical imperative which clears the mind of hesitation and uncertainty. And yet, even here, Haag was working upon what already existed. The pond was there when he first saw the site, but the owners were not very happy with it:

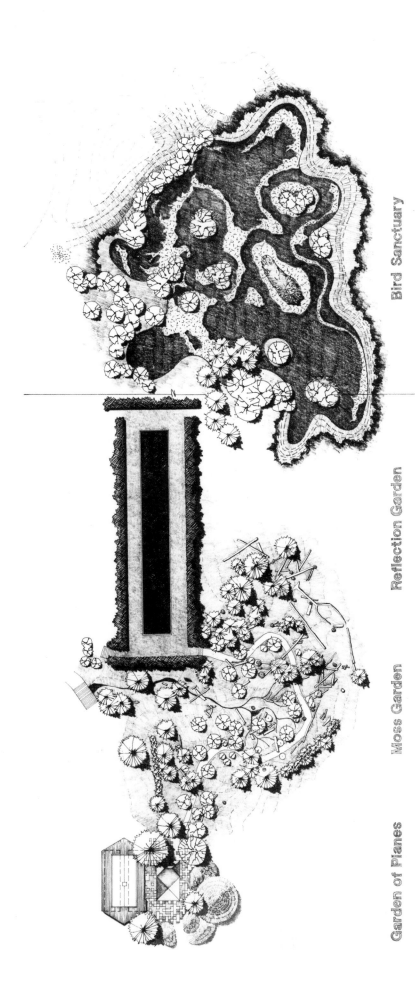

Garden of Planes   Moss Garden   Reflection Garden   Bird Sanctuary

A SERIES OF GARDENS · THE BLOEDEL RESERVE
RICHARD HAAG · LANDSCAPE ARCHITECT · 1985

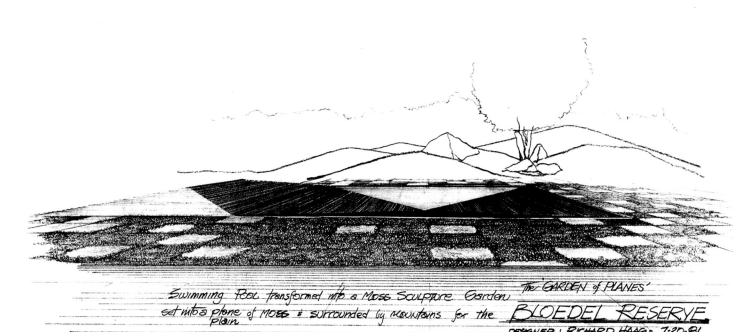

Swimming Pool transformed into a Moss Sculpture Garden The "GARDEN of PLANES"
set into a plane of MOSS & surrounded by mountains for the BLOEDEL RESERVE
plain
DESIGNER: RICHARD HAAG - 7-20-81

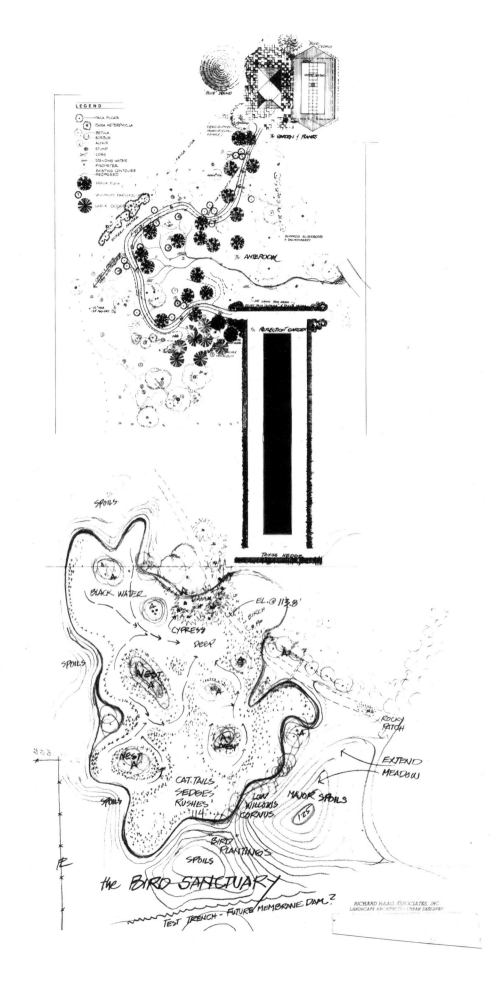

现场的时候，水池就在那里，但是业主并不喜欢它：冰冷黑暗的水体与布洛德尔心中梦想的"欧式风格"公园格格不入。然而，哈格把这个巨大水体看作镜子，然后用红豆杉树篱创造适当的框架，以聚焦它的效果，包装它的反射。只是小小的调整就创造出让人震惊的效果！

穿过黑森林走一会儿，就来到了"鸟类乐园"。这位来自西雅图的景观建筑师通过改造灌溉水塘和增加新的宽阔水面来创造景观，两个湖相隔不远，在设计的时候听取了鸟类学家的意见，以便于能够容纳尽可能多的鸟类，包括本地鸟类和迁徙鸟类。一些小岛点缀在宽阔的水面上，布局经过精心的设计，以使它们相互掩映，随着移动时视点的改变，不断有新的小岛进入视野。哈格醉心于这种视野的游戏，之所以这么做，是专门设计给更有心的参观者，阅读现实的额外发现。哈格旨在将这片自然的空间塑造为完全排除人类干扰而尊重自然的区域。事实上，这片区域和园区的其他部分没有明确的联系；目的是让人们在这里能够——保持一定距离——观察野生动物。然而，现在那些负责公园的人员，在湖泊周围修起了步行道，这画蛇添足的一举，完全扭曲了哈格的规划理念。更有甚者，现在的管理者做出了令人遗憾的决定：把"平面花园"改造为将来令人为难的禅宗庭园。哈格最初设计中的那些富有远见卓识的创意将荡然无存。

its cold, dark water seemed to have little to do with the dream of a 'European-style' park that the Bloedels had in mind. Haag, however, saw this expanse of water as a mirror, and then used the yew hedge to create an apposite frame to concentrate its effect and encase its reflections. Just a few adjustments to create a stunning result!

A short walk through the dark forest one comes to the Bird Sanctuary, which the Seattle landscape architect created by transforming an irrigation basin and adding a new expanse of water. The two 'lakes' are a short distance apart and were designed with the help of an ornithologist so as to be able to serve the largest possible number of birds – both migratory and resident species. Small islets dot the expanse of water, their distribution carefully designed so that they conceal one another, with new islands becoming visible as one shifts one's point of view. It is the sort of play upon perspectives in which Haag delights, designed specifically to provide the more attentive visitor with an additional key to the reading of reality. Haag intended this 'natural' space as a zone in which Nature is respected to the total preclusion of man. In fact, there is no precise link between this and the other areas of the estate; the aim was that one should here be able – from a distance – to observe the wildlife. However, those now responsible for the park have chosen to surround the lakes with pathways, an addition that totally traduces the idea behind Haag's plans. Furthermore, the present administration also took the regrettable decision to turn the Garden of Planes into an embarrassing would-be Zen garden, which has none of the visionary acuteness of Richard Haag's original designs.

12.

斯塔米诊所

北纬：47°50′44.07″
西经：122°16′30.05″

地点：林恩伍德市，华盛顿州，西南169号大街
时间：1962年
状况：已建成
项目经理：理查德·哈格
设计团队：理查德·哈格事务所（弗兰克·杰姆斯）
项目委托方：Dr. Arthur Stamey
摄影师：courtesy RHA

在20世纪60年代初期，哈格移居到西雅图之后，承接的项目中既有私人委托的又有公共部门委托的。作为单一方式的互补，他通过注重形式的品质在两个领域都表现出特色。林恩伍德是西雅图以北几英里的一个小镇，为斯塔米医生设计的诊所就在这里。哈格再创造了典型的日本传统庭园建筑形式，跨越了抽象而组织的适应纯功能需要的一种模式。对于没有任何强有力明显特征的、坐落在城市肌理边缘区的建筑，如何赋予一些重要特征，这个项目给出了答案。院落的围墙是由石头堆起来的，那流畅的曲线是从数学中的抛物线和双曲线得到灵感的。实际上，与其说这是墙，不如说是雕塑，由石头结合在一起，没有任何明显使用的混凝土痕迹。如此，哈格赋予这个既没有历史也没特色的地方以具有标志性的形象。不知何时沿着河床散布的石头，是自然界呈现混沌的最好表达，变成了完美和卓越的象征，与那时包围着小牙科诊所的、具有不可驯服外表的森林形成强烈对比。这些起伏的河石不仅能够遮挡房屋的入口，限定停车场的布局，还能把道路上的车流挡在外面。禅宗概念里，石头是力量的化身，人类是无力的；这充分说明在自然的话语中，上下关系是颠倒的。这里石头的布局是按照精确的设计来摆放的。它们是人类理性的表达，与周围杂乱无章的野生自然环境对比鲜明，也成为解读自然的度量。从几何形式中得到灵感，这个有序与无序的清晰对置点亮了万物本源的美丽。

At the beginning of the 60s, just after his move to Seattle, Haag divided his time between private projects and those for the public sector. Complementary aspects of a single approach, his work in both areas was characterised by attention to the quality of form. Lynnwood is a small town a few miles north of Seattle, and in designing a clinic there for Dr. Stamey, Haag 'reinvented' a form that was typical of historical Japanese garden architecture, moving beyond abstraction to organise this model to meet purely functional needs. The project resolved the problem of how to give some importance of character to architecture that was without any powerful distinguishing features and stood within an area that (at the time) was on the margins of the urban fabric. The enclosure around the property thus becomes a system of stone walls whose clean curves are inspired by mathematical parabolas and hyperbolas. Indeed, more than walls, these are sculptures, made of stones fitted together without any apparent use of concrete. As a result, Haag endowed a place with no specific history or character with an iconic presence. The stones which, when scattered along a river bed, are the highest mineral expression of the chaos that can be present in Nature, here became symbols of perfection and transcendence, standing in sharp contrast to the untamed appearance of the forest which (at the time) still surrounded the small dental surgery. These oscillating forms in river stone also serve to shelter the building's entrance point; to regulate the layout of the car park; and to shut out the flow of traffic along the road. The Zen conception of stones as embodiments of strength, of man's inability to fully interpret that which Nature is saying, is here turned upside down. The stones are organised in accordance with a precise design. An expression of human rationality, they stand in clear contrast to the untamed natural setting; they become the measure on which interpretation is based. Inspired by geometrical forms, this clear contraposition serves to highlight the original beauty of the universe.

13.

布鲁姆住宅

地点：阿斯彭，科罗拉多州
时间：1988～1990年
状况：已建成
项目设计：戴维·芬霍尔姆
项目经理：理查德·哈格
设计团队：理查德·哈格事务所（罗伯特·弗雷）
项目委托方：Neil + Barbara Bluhm
主要承建方：Hansen Construction Inc; Arrow J Landscape
摄影师：courtesy RHA
场地面积：4.5公顷（11英亩）

一个新近的项目，落基山景观给予了理查德·哈格机会，将他成熟作品的各种主题糅合在一起。设计的目的不是为了强调项目本身，而是让它成为既有环境中的一部分。在色彩的选择上，石头元素与泥土和芳草巧妙地融合。它们在庭园以色彩缤纷而优雅精致的自然布局方式蔓延，作为远处高山景观的补充而存在。清澈的天空和冷杉树通过在泳池尽头入口处的"舞台拱门"而被框景，成了令人称赞和引人注目的艺术作品。剧场的"隔间"由被桦树包围的大型院落组成，形成了以落基山坡为背景点缀着奇形怪状石头的幕帘。在这个大型别墅周边，哈格布置了水道形成的溪流。溪流仿佛叙述着窃窃私语的童话，并吸引和引导着附近真正的源头：一条山里的河流因而有机会变得更为纯洁。

布鲁姆住宅庭园是一个为富有客户设计的奢侈项目。然而，理查德·哈格利用这种富裕去聚焦简单事物的美。规划的植被既不奢侈，也不陌生，与科罗拉多的山峰景观和谐共生。游泳池得益于天空的颜色，溪流是对冬季覆盖在山脉上积雪的赞誉。各种人造景观的并列创造了一种组合：外观上自然但实际上却是一个技术、建筑、植物、种植选择之间良好平衡的产物。又一次，理查德·哈格仅采用大自然提供的素材，通过新颖的设计满足了客户的需要。各种空间和区域的选择与组织，将景观组织成整体的联系，项目蕴含着另一番更为深奥解读的事实，这些都是哈格艺术的典型特征。

A late project, this opportunity to interpret the Rocky Mountain landscape gave Richard Haag the chance to bring together all the themes of his mature work, within a design that did not aim to measure itself against the existing environment but rather to become part of it. Chosen for their colour, the stone components blend in with both earth and grass. As they extend through the garden, the paths unfold through a natural setting which is colourful and refined; which stands as a complement to the alpine landscape in the distance. The clear sky and the fir trees are seen through the 'proscenium arch' of the portal that stands at the end of the swimming pool; thus framed, they become a work of art to be admired and contemplated. The 'stalls' of this theatre comprise the large patio that is enclosed by birch trees, forming a curtain against the mountain slope dotted with erratic stones. Around the large villa, Haag has laid out a watercourse to form a stream that recounts a whispered fairytale; captured and channelled near its very source, a mountain river thus becomes an opportunity for the re-creation of innocence.

The garden of the Bluhm residence is a rich project for a rich client. Richard Haag, however, exploits this opulence to focus upon the beauty of simple things. The vegetation is neither extravagant nor alien to its setting; it lives in symbiosis with the landscape of the Colorado peaks. The swimming pool pays tribute to the colour of the sky, and the stream is a recognition of the snows that cover the mountains in wintertime. The various man-made juxtapositions create a composition that is apparently naïve but is actually the product of a fine balance between technology, architecture, botany and plant choices. Once again, Richard Haag plays with materials that Nature provides in order to meet the requirements of the client without ever having to settle for the banal. The selection and organisation of the various spaces and areas; the relation with the landscape as a whole; the fact that the project is open to a second, more profound, reading – all of these are characteristic features of Haag's art.

14. 西雅图中心

北纬：47°37′21.05″

西经：122°21′7.27″

地点：西雅图，华盛顿州，哈里森大街 305 号
时间：1962 年（第一期）；1977 年（修复）
状况：已建成
项目设计：鲍尔·西提
项目经理：理查德·哈格
设计团队：理查德·哈格事务所（格兰特·琼斯，罗伯特·汉娜）
项目委托方：City of Seattle
摄影师：courtesy RHA
场地面积：30 公顷（74 英亩）

西雅图市确切的市中心很难确定，但是它的标志却毫无疑问。605 英尺高的标志性建筑"西雅图中心空间指针"，位于 1962 年世界博览会的场地（二战后首次在美国举办）。正如这次盛会的主题"生活在 21 世纪"和我们现在所知的一样，博览会呈现了未来的图景，展示了在各个领域因技术发展而拥有可能性的巨大而直率的信心。展会结束后，留下来的只有指针般的高塔、单轨铁路和绿地，理查德·哈格的设计是将各种各样的空间连接在一起。被编织的景观包裹着像歌剧院和太平洋科学中心这些功能不同且空间巨大的设施，在总计 74 英亩的范围里，创造了一个序列的公共花园。不同的空间被组织成相互开放、可识别清晰的小空间，并成为统一整体的一部分。喷泉庭园、国旗广场、被轻微缓坡和绿茵覆盖的地形包裹着的圆形剧场、休闲树林（为孩子们和富有童趣的人们而设置的），这些都是景观提供了基本构架的统一整体的组成部分。项目最基本的考虑是将分离的元素编织在一起，理查德·哈格选择了一种简单而易陈述的语言。作为一位智慧而苛求的景观建筑师，他对使用一些让人难忘的奇特造型没有兴趣，而是认为将西雅图中心作为一个令人愉悦的背景音乐保留在记忆里更为重要。这一系列空间正是让人们去体验的，而不是企图去统治的。在 1997 年，哈格被委托来重新解读并修缮他的项目，以适应公众新的需求和中心附近出现的新建筑。甚至在"空间指针"附近新建的弗兰克·O·盖里的华而不实金属感极强的"体验音乐项目"之后，理查德·哈格轻描淡写的一排树和宽阔起伏的草地都完好无缺地保留着。

The actual city centre of Seattle is rather hard to identify, but there is no doubt about its symbol: the 605-feet Space Needle of the Seattle Center, the location of the 1962 World Expo (the first to be held in the post-war USA). The event took as its title 'Living in the 21st Century' and – we know now – presented an image of the future which showed enormous and ingenuous trust in the possibilities that technological development would make available in numerous fields. All that remains now is the Needle itself, the monorail and the green area which – after the Exposition itself had come to an end – Richard Haag designed to link together the various spaces. The landscape is woven around such different and widely-spaced facilities as the Opera House and the Pacific Science Center, creating a sequence of public gardens that cover a total area of 74 acres. The different spaces are organised as rooms that open off each other, distinct and yet clearly recognizable as part of a single whole. The fountain court, the flag piazza, the gently-sloping grass-covered terrain around the amphitheatre, the entertainment wood (intended for children and for everyone who is young at heart) – all these are the components of a single composition in which landscape provides the basic structure. In a project that was primarily concerned with weaving together separate elements, Richard Haag opted for a simple, understated language. As an intelligent and demanding landscape architect, he was not interested in coming up with some striking feature which would get him remembered; it was more important that the Seattle Center should remain in the memory like a pleasant background noise. This is a series of spaces that offer themselves up to lived experience rather striving to dominant it. In 1997 Haag was commissioned to 're-read' and restore his project, taking into account the new needs of the public and the new buildings that had arisen near the Center. And even after the creation near the Space Needle of Frank O. Gehry's exuberant metallic Experience Music Project (2000), Richard Haag's understated strand of trees and wide undulating lawns remains intact.

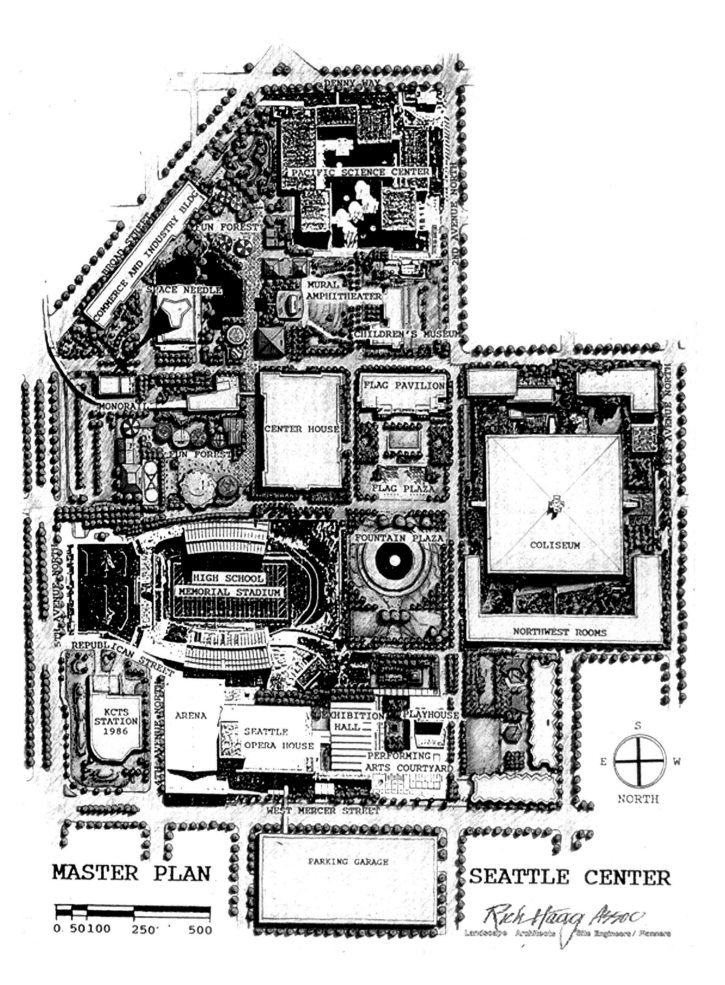

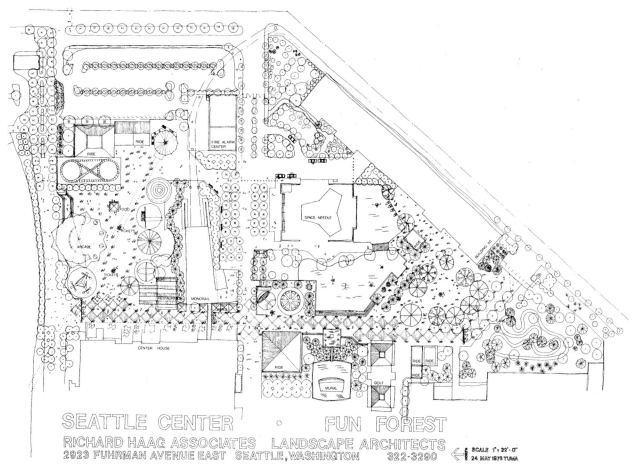

15.

拉维莱特公园竞赛

北纬：48°53′34.66″
西经：2°23′28.06″

地点：巴黎，法国
时间：1982 年
状况：未建成
项目经理：理查德·哈格
设计团队：理查德·哈格事务所
艺术家：Kenn Rupard

我们都知道，1982 年为巴黎拉维莱特公园举办设计竞赛和 1987 年项目实施之后发生的事情。不经意间，这个公园成为区别于现代建筑的当代设计。在与哲学家雅克·德里达的合作下，瑞士建筑师伯纳德·屈米（当时几乎是新手）的获胜作品应用了结构主义理论并实践了它。整个设计用植物作为一种"填充物"，来联系广泛空间散布着的建筑的特征，这两种成分在不同平面上起作用。拉维莱特被景观建筑师们（当时很多人参与了这个竞赛）设计的方案已经被记录在案。理查德·哈格自己也提出了方案，方案中自然性成为整个 35 公顷老屠宰场全部再开发的统领特征。地形和植物特征使空间充满活力：起伏的地表创造了新的地平线，与宽阔的中央林荫大道成为一体。林荫大道穿过乌尔克水渠，连接公园的两部分。一排排的树木旨在建造随着时间流逝而变化的特征，通过季节变更的塑造，这些元素把公园里现存建筑与新建的建筑带来的想象连接起来而产生新的理解。和一些乌托邦式的对一般主题的探索不同，这个项目提供了对具体要求可能的回应。设计线条简洁，展现在宽广的尺度上，空间以一种线性的、符合逻辑的序列组织起来。从南部有序的城市肌理穿过庆祝生命的欢乐的明显无序，然后穿越乌尔克水渠，公园的肌理扩散开来，拥抱向巴黎的郊区。就像在他所有的项目里一样，哈格运用高度、梯度、尺度和细节的层次，以创造一个完整的公共空间。这是一个自由而坦率的公园，为现实的人们而创造，与他们和谐共存。

We all know what happened after the design competition (in 1982) for the Paris Parc de la Villette and the completion of the project in 1987. Quite unintentionally, the park has become that which discriminates contemporary from modern architecture. Developed, in collaboration with the philosopher Jacques Derrida, by the Swiss architect Bernard Tschumi (then almost a newcomer), the winning project applied deconstructionist theories to reality itself. The entire design used vegetation as a sort of 'filler' to link the widely-spaced architectural features, the two components working at parallel levels. La Villette as designed by landscape architects (many of whom took part in the design competition) has however remained on paper. Richard Haag himself proposed a solution in which nature became the feature around which hinged the entire redevelopment of the 35-hectare site of the old slaughter-house. Features of terrain and vegetation served to bring spatial volumes alive: the rise and fall of surfaces created new horizons which integrate with the wide central avenue that passes under the Ourcq canal joining the two parts of the park. Lines of trees were intended as architectural features that changed with the passage of time; modelled by the changes of season, these volumes would serve to create new perceptives that linked together the existing buildings and those envisaged for creation within the park. Far from being some utopian exploration of a general theme, this project offered a possible response to specific requirements. The lines of the design are simple and unfold on an ample scale, with the spaces organised in a linear and logical sequence. From the ordered urban fabric in the south one passes to the apparent disorder that celebrates the joy of life; and thence – beyond the Ourcq canal – the fabric of the park spreads out to embrace the Parisian suburbs. As in all his projects, Haag here exploits hierarchies of height, gradient, size and detail in order to create a 'complete' public space. Free and forthright, his is a park created for – and in tune with – actual people.

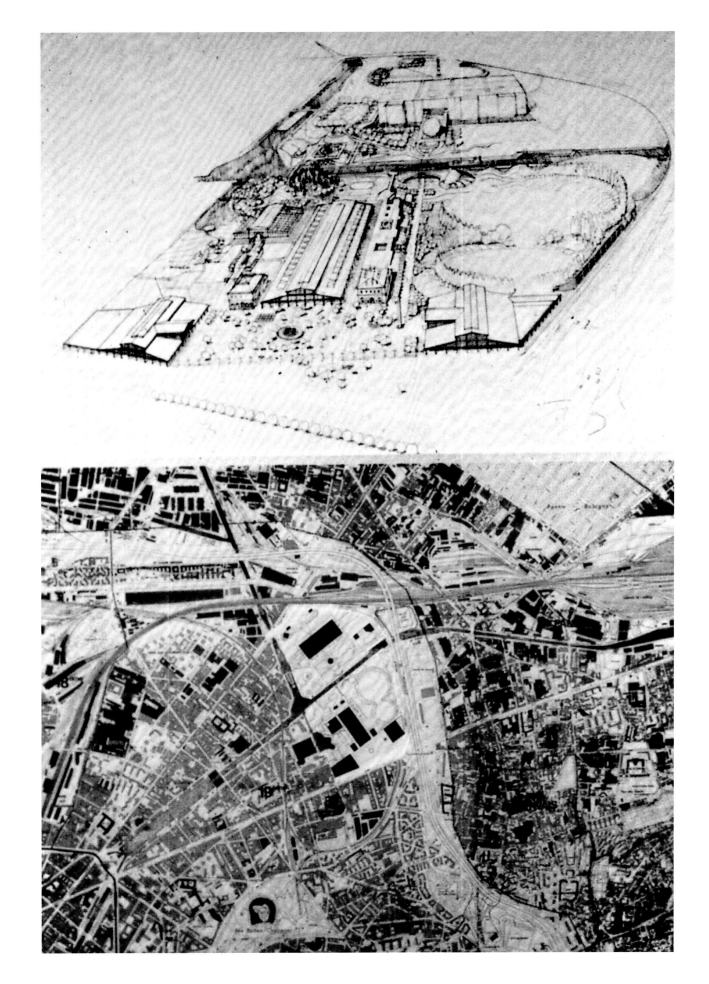

16. 吉尔曼村庄／吉尔曼林荫大道

北纬：47°32′25.66″
西经：122°2′25.50″

地点：伊萨夸市，华盛顿州，吉尔曼林荫大道西北
时间：1983～1985 年
状况：已建成
项目设计：贝利斯建筑师
项目经理：理查德·哈格
设计团队：理查德·哈格事务所（史蒂文·G·雷，罗伯特·汉娜）
项目委托方：City of Issaquah
工程设计：Myron Anderson
摄影师：courtesy RHA, Luca M.F. Fabris
场地面积：林荫道 2.4 公里（1.5 英里）；村庄 1.2 公顷（3 英亩）
建造费用：林荫道 250 万美元

伊萨夸是一个兴建于 19 世纪末的矿业城镇，它的名字来源于当地前哥伦比亚土著语言的单词"鸟的歌声"。1970 年代，理查德·哈格说服了投资人马文·莫尔和他的妻子露丝买下长期被遗弃的小木屋并把它们转变成一个商业中心，叫做"人的尺度"。这个主意获得了成功，理查德·哈格事务所进行了场地规划和绿地设计，吉尔曼村庄作为步行者的绿洲被保留了，在这里人们能够以悠闲的步调购物，从一个场所移动到另一个场所(即从一个商店到另一个商店)。在 1986 年，伊萨夸市当地政府委托哈格设计入城的主要道路——主要为了提升从高速公路到城镇中心的景观形象。哈格因而拥有了探索实践他激情想法之一"食材"景观的机会。概念很简单：用树木和灌木来限定道路，使用不同颜色的叶子、花卉和果实（总的来说是供给当地社区的来源）来设计景观。吉尔曼景观大道项目开始于大量数据的收集，以挑选适应当地又足够强壮，能抵御长期交通影响的植物品种；同时，各种树种在不同时期开花的周期经过仔细的计算，以保证果实的收获持续整个夏季甚至更长。

限定 4 车道高速公路的"骨架"结构由一行行的白杨树构成，白杨是当地经常被用于标识领地界线和形成防风林的树种。其他的树种——例如枫树——被用来作为背景衬托种在路边上的梅树、苹果树和梨树。不幸的是，"食材"景观理论的整个实践，遭遇了忽视规则和忽视关照，我们的社会经常认为这些是过时的。就他这方面而言，理查德·哈格做出了自己的说明，一个人能在任何地方创造出"食物链"，只要他下定决心，在自家种植广泛选择的果树——养育来自沿太平洋的美国西北海岸的本土植物——以取悦自己的家人、当地的小鸟或是附近学校的孩子们。

Issaquah is a mining town founded at the end of the nineteenth century; its name comes from a term for 'birdsong' in the language of the pre-Columbian inhabitants of the region. In the 1970s Richard Haag convinced the investor Marvin Mohl and his wife Ruth to buy the long-abandoned small wooden houses and transform them into a commercial centre 'to the measure of man'. The idea was a success, and Gilman Village – the site plan and the green areas of which were designed by Richard Haag Associates – still survives as an oasis for pedestrians; as a place where one can shop at a leisurely pace, moving from house to house (that is, store to store). In 1986 the local administration of Issaquah commissioned Haag to design the main approach to the town – and primarily to improve the appearance of the highway that leads towards the town centre. Haag thus had the opportunity to explore one of his passions: the 'edible' landscape. The concept was simple: defined the road with trees and shrubs so as to design the landscape using the different colours of foliage, blossoms and fruit (a source of nutrition for the community as a whole). The Gilman Boulevard project started with the collection of a large amount of data regarding the species best suited to the habitat and robust enough to withstand the long-term effects of traffic; at the same time, the different periods in which the various trees blossomed was carefully calculated, so as to guarantee a fruit crop throughout the summer and beyond. The 'skeletal' structure delimiting the four-lane highway was provided by lines of poplars, trees which have always been used in this region to mark property boundaries and to create windbreaks. Other species of tree – for example, maples – then served as background support to the plum, apple and pear trees laid out along the sides of the road. The complete application of the theory of an 'edible landscape', unfortunately, comes up against the neglect of rules and gestures of care and attention which our society often considers to be out-of-date. For his part, Richard Haag has demonstrated that one can create a 'food-chain' wherever one sets one's mind to it by surrounding his own house with a wide selection of the berry-bearing shrubs indigenous to the north-west Pacific coast-to the delight of his own family, the local birds and the children at a nearby school.

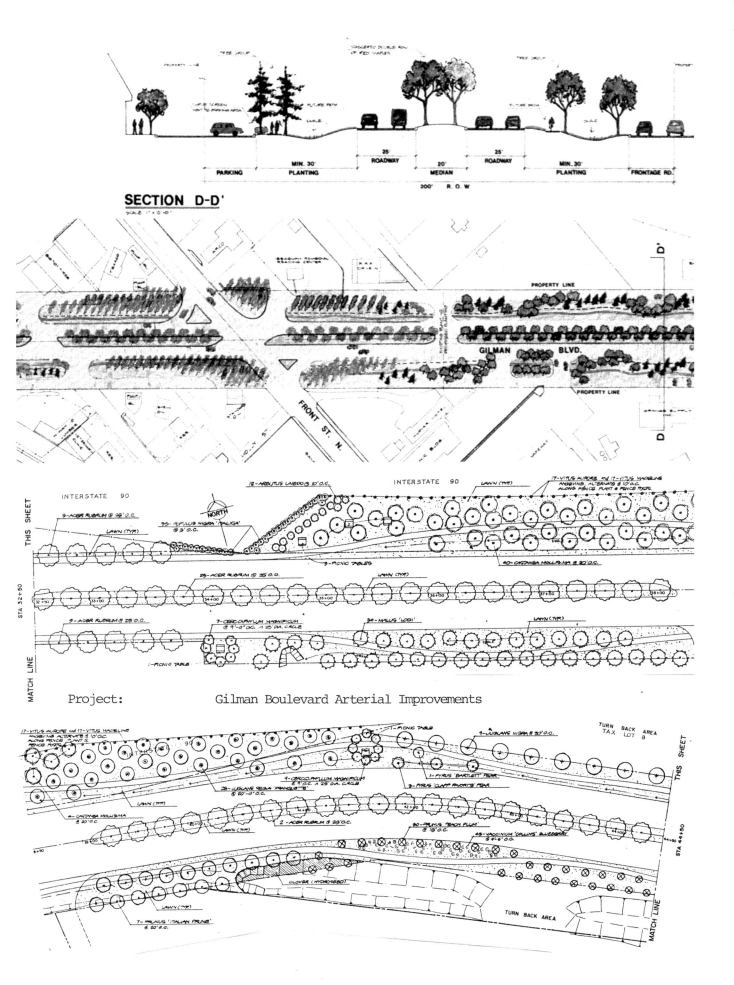

17.

亨利·M·杰克逊联邦大厦

北纬：47°36'17.55"

西经：122°20'6.70"

地点：西雅图市，华盛顿州，第二大街915号
时间：1970～1974年
状况：已建成
项目设计：弗雷德·巴塞蒂及其合伙人公司
项目经理：理查德·哈格
设计团队：理查德·哈格事务所（罗伯特·汉娜）
项目委托方：U.S. Government
工程设计：John Graham Associates
摄影师：courtesy RHA, Luca M.F. Fabris
艺术家：Isamu Noguchi
场地面积：0.6公顷（1.5英亩）

弗雷德·巴塞蒂及其合伙人公司设计的490英尺高且平面方正的摩天大楼是20世纪70年代早期在西雅图市中心核心部的许多条道路的地标。一方面，基地原址被"复兴罗马主义"的伯克建筑所占据。该建筑完成于1891年，由亨利·H·理查德森设计。它代表了1889年可怕的大火后，城市中心区重建采用的众多折中主义建筑语言的一种。部分是为了安抚反对拆除已经成为西雅图城市肌理历史组成部分的抗议者，着手设计新建筑的建筑师决定部分保留老建筑，使它们与摩天大楼周边规划的一个公共空间相协调。受巴塞蒂的委托，理查德·哈格从这里出场了，提出一些恰当而不朽的方案，以解决第一大街与第二大街处于不同标高带来的问题，还需要解决与老建筑的保留遗迹相协调的问题。哈格的构想是在罗马式建筑中运用西班牙台阶。然而，哈格设想的解决方案，在某些方面甚至比建筑师的要求更出色，它还重视考虑场所缺乏深度和遵照项目要求建造一个安全而可控的通道。在摩天大楼的南面，开放空间在第二大街的部分，协调于老伯克建筑入口处的现存拱门，呼应野口勇的雕塑"时间景观"。从这里下去是一个关节相连的台阶系统，红色的河流在台阶护栏之间翻滚，成功地容纳了它多变的路径。所使用的材料很简单——砖、赤土色瓷砖和混凝土、上好的砾石和锃亮的不锈钢扶手——为植物创造了一个限定的背景。在高层建筑、台阶系统、平台与休息空间之间，尺度上有着深思熟虑的不一致，但又存在联系。视线是清晰的，提供远景和眺望，强调了"意大利的"或者说是"罗马的"设计精神。触摸光线意味着对巴洛克本源的哈格式的再解读，充分反映于他那个时代"选择性"的流行文化中。优雅的隐喻创造了公共社会空间，回避了任何形式僵化的围合。这个革命性的设计赢得了1981年美国景观建筑协会的荣誉奖。

Standing almost 490-feet high, the square-floorplan skyscraper designed by Fred Bassetti & Company for the heart of downtown Seattle in the early 1970s is emblematic in a number of ways; for one thing, the original site had been occupied by the 'Revival Romanesque' Burke Building which, completed in 1891 to designs by Henry Hobson Richardson, had exemplified one of the many eclectic architectural languages adopted when rebuilding the city centre after the terrible 1889 fire. Partly in order to placate the protests raised against demolition of what had become a historic part of Seattle's urban fabric, the architects of the new structure decided to keep parts of the old building and incorporate them in the layout of a public space to be created around the skyscraper. This is where Richard Haag came in, commissioned by Bassetti to come up with some suitably monumental resolution to the problem posed by the difference in level between 1st and 2nd Street – a resolution that should also incorporate the preserved remnants of the old structure. Haag's idea was to take the Spanish Steps in Rome as a model. However, the solution devised by Haag is, in some ways, even better, given that it takes into account both the lack of depth in the site and the need to comply with such project requirements as safety and controlled access. On the south side of the skyscraper, the open area giving onto Second Street frames the extant arch of the entrance to the old Burke Building, which is echoed by Isami Noguchi's sculpture 'Landscape of Time'. From here descends an articulated system of steps, a river of red tumbling between parapets that just manage to contain its shifting course. The materials used are simply – brick, terracotta tiles and concrete, with fine gravel and handrails of burnished steel – creating a defined setting for the plants themselves. There is a deliberate incongruence of scale between the high building and this system of steps, terraces and resting-places – but there is a relation. The sightlines are clear, providing perspectives and outlooks that underline the 'Italian' – or, better, 'Roman' – inspiration of the design. Lightness of touch means that Haag's re-reading of the baroque original fully reflects the 'alternative' Pop culture of his own day. Refined quotation serves to create public social spaces, eschewing any type of fixed enclosure. This was a revolutionary design that in 1981 won the American Society of Landscape Architecture's Honor Award.

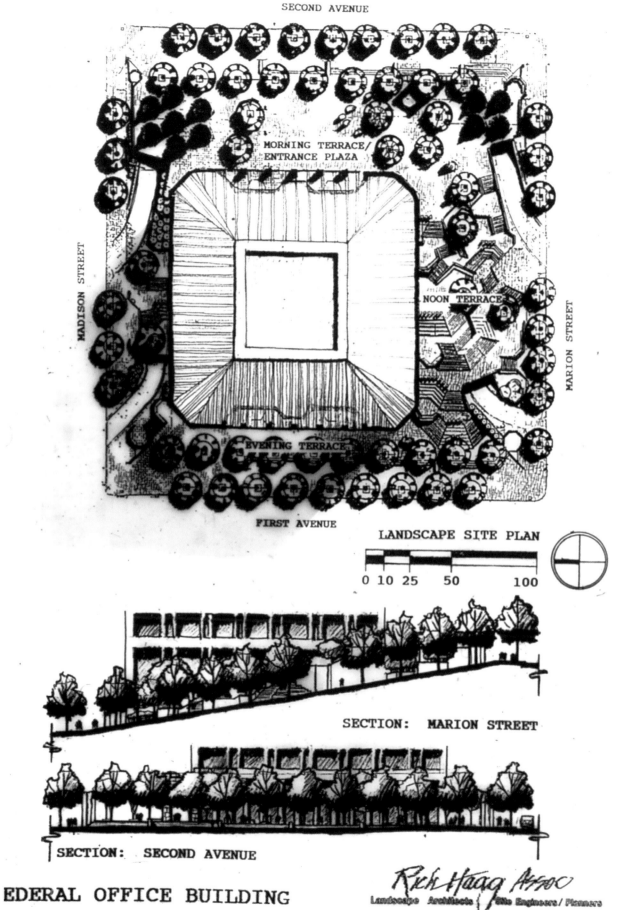

18.

梅里尔庭院式联排住宅

地点：西雅图市，华盛顿州，哈佛大街东901号
时间：1986年
状况：已建成
项目设计：伊本·内尔森事务所
项目经理：理查德·哈格
设计团队：理查德·哈格事务所（布拉德·黑川）
项目委托方：Merrill Court Association
工程设计：Victor O. Gray, TML/Stern Engineers, Travis Fitzmaurice Associates
主要承建方：Eberharter Construction Group
摄影师：Mary Randlett, Luca M.F. Fabris
场地面积：0.4公顷（1英亩）

要想领会这个项目的特别之处，就必须理解国会山代表西雅图的什么。这个稍有坡度的小山形成尤宁湖和华盛顿湖之间的半岛，已经作为城市富裕阶层居住区超过一个世纪。它栽满树木的街道为住宅提供了华丽而不炫耀的宽阔背景。这里的城市肌理由独栋的家庭住宅组成，因此公寓街区远远不合通常的规范。这就是为什么建筑师伊本·内尔森选择了一个由20世纪早期"田园城市"所激发的设计，并将美国的场地背景与英式风格结合起来的原因。

颠覆了国会山不成文的规则，理查德·哈格设计了一个半开放空间，使其作为一个庭园，既保护了建筑物又建立了与街道的直接联系。在哈佛林荫道东和东阿罗哈大街拐角处的树篱后，开放出一片缓坡草地，针叶树组成的树丛在基地已经存在，部分树长在表面贴砖的L形建筑面前相得益彰。一个稍有弧度的砖砌小径伸向公寓，穿过小型的私人花园或露台。通过使用这种策略，哈格使得整个绿色空间看起来像一个连续的整体，并给作为背景的建筑体一种深远感。车库位于公寓的下方，从街上可以直接进入。景观建筑师利用建筑内的不同层高，形成一系列的"空中花园"。受英式台地住宅模式的启发，方案设计了连通10家住宅表面的绿墙。在用绿化丰富可获得的空间，促成多样化小气候形成的过程中，哈格仔细地开发利用当地的植物物种。所有的美都是通过并排连通建筑物而释放，并得益于水平和垂直面上色彩的对比和相互作用。

To grasp the exceptional character of this project one has to understand what Capitol Hill represents for Seattle. Forming a peninsula between Union Lake and Washington Lake, this slight hill has for more than a century been a residential area for the city's wealthy, its tree-lined streets providing an ample setting for homes that are luxurious but not showy. The urban fabric here is made up of detached family houses, hence an apartment block was far from fitting in with the norm. This is why the architect, Ibsen Nelsen, opted for a design inspired by the 'garden cities' of the early twentieth century, combining an American setting with hints that are English in tone.

Subverting the unwritten rules of Capitol Hill, Richard Haag designed what is a semi-public space, a courtyard that both protects the building and establishes a direct relation with the street. Behind the hedge on this corner between Harvard Avenue East and East Aloha Street opens out a slightly sloping lawn; the bosk of conifers trees was already present on the site and now stands in proportion to the L-shaped building faced with brick. A slightly curving brick pathway leads towards the apartments, which are approached through small private patio/gardens. By adopting this strategy, Haag makes the entire green space appear as a single continuous whole, giving a sense of depth to the setting of the architectural complex. The garages are located under the apartments and are accessible directly from the street. The landscape architect exploits differences in level within the structure to form a series of 'hanging gardens', which create a wall of greenery across the facades of the ten homes, whose juxtaposition is inspired by the English model of terraced housing. In his use of greenery to 'multiple' the available space and generate variation in micro-climate, Haag deliberately employs local varieties of plant. Their full beauty is thrown into relief by juxtaposition with the buildings themselves – thanks both to colour contrasts and the interplay between vertical and horizontal planes.

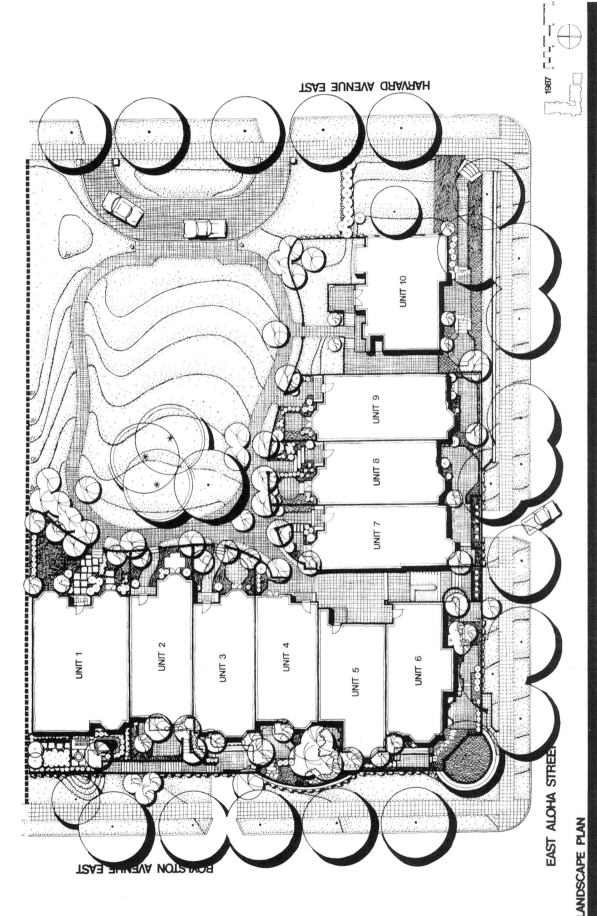

华盛顿州生态总部设计竞赛

北纬：47°2′52.24″

西经：122°48′37.71″

地点：莱西市，华盛顿州，德斯蒙德车道东南 300 号
时间：1986 年
状况：已建成
项目设计：汤普森，韦沃达及合伙人公司
项目经理：理查德·哈格
设计团队：理查德·哈格事务所（罗伯特·弗雷）
项目委托方：Washington State Department of Ecology
工程设计：Rosewater Engineering
主要承建方：Hoffman Construction
场地面积：11 公顷（28 英亩）

在位于莱西市的华盛顿州生态部门新总部的三轮设计竞赛中，最后拔得头筹的是一个以"生态概念"为核心的设计方案，理查德·哈格在场地中融入了历史和自然环境的元素。该方案由内德·韦沃达的汤普森工作室和波兰的韦沃达伙伴公司负责设计，建筑本身方案取自典型的俄勒冈州工作室：没有一味追随潮流，而是保持了古典功能主义倡导的以服务设施为导向的布局形式。因此，景观设计师不得不去做融合建筑本身的设计，在绿地中混合建筑物刚性的表皮，以跨越城市与乡村的鸿沟。哈格与该项目城市规划方面的人员协作，选择围合一个典型大学校园的开敞空间。这个空间被一排排的树划分为开放的"小空间"，创造出宽阔的草坪、停车场和道路入口空间。附近的圣马丁修道院和与其同名的圣马丁学院同时修建于 1901 年，它们提供了一个历史的参考点。哈格设计增加的针叶树，从建筑边缘一直延续到树林，成为那里突出的特征。

事实上，哈格的作品是被时间限定的，随着设施本身的发展能够生长和变化。建筑两翼之间的大型围合庭园、水池和水池中利用影子测量时间的大型日晷，这些在景观规划中很少是永恒固定的样子。而剩下的元素在原地也许会存在几个世纪或者被迅速地改变。室外空间的特点没有刻意限定地被仔细保留。没有绝对的边界，意味着场所保留了很大程度上的灵活性。

The three-phase competition for the new headquarters of the Washington State Department of Ecology in Lacey resulted in a design that epitomises the very notion of 'the ecological', with Richard Haag engaging with both the history and natural setting of the site. Designed by Ned Vaivoda of the Studio Thompson, Vaivoda & Associates (Portland), the building itself follows a schema that is typical of the Oregon studio: with no concession to fashion, it remains faithful to the classic functionalist layout of service industry facilities. It was therefore the landscape architect who had to absorb the architecture within its setting, blending the solid presence of the structure within a green area that still straddles the divide between the rural and the urban. Haag collaborated on the urban-planning aspects of the project, opting for the inclusion of the open areas typical of a university campus; divided by lines of trees, the open-air 'rooms' thus created expanses of lawn, car parks and access roads. The vicinity of St. Martin's Abbey and the College of the same name (both built in 1901) provided a historical reference point, exploited by Haag to 'structure' the addition of a conifer wood; this expands.from the edge of the property into the forestland that is still the prominent feature of the area.

In effect, Haag's project is to be defined by time; it can grow and change as the facilities themselves develop. The large enclosed gardens between the wings of the building, the water pools and the large sundial whose shadow measures time upon the surface of the water – these are the few permanently fixed aspects of the landscape layout. The rest could remain in place for centuries or could be altered very quickly. The character of the external spaces has deliberately been left without strong definition; this absence of categorical boundaries means the site has maintained a large degree of flexibility.

20.

美国驻葡萄牙里斯本的大使馆和领事馆

北纬：38°44′37.75″

西经：9°9′44.66″

地点：里斯本，葡萄牙，爱文尼达·达斯·佛凯斯·阿玛达斯
时间：1975～1983年
状况：已建成
项目设计：弗雷德·巴塞蒂
项目经理：理查德·哈格
设计团队：理查德·哈格事务所（戴勒·丹尼斯）
项目委托方：U.S General Services
摄影师：courtesy RHA
场地面积：5公顷（12英亩）

作为理查德·哈格在欧洲的一个项目，里斯本美国大使馆项目涉及现有场地的再开发和重新定位。建筑师弗雷德·巴塞蒂利用该场地和老贵族别墅轻微抬起的场地之间的不同高差，创造了一个建筑意味的小瀑布，以包容并协调大使馆和领事馆多样化的服务设施。它是一个简单而常规的布置，把巴塞蒂典型设计中清晰合理的外形与州政府规范的安全要求相结合。在这个围合而内向的环境里，理查德·哈格的设计将大家的视点从已建成的建筑中，转向了大使馆场地中央的大片地中海松林上。同时，设计增加了建筑后方的私密性，从街上看去，建筑几乎完全被隐藏起来。这一大片树木也变成了一种宜人的设施，用来建立内部和外部空间的交流。得益于传统屋顶的红瓦和稠密墨绿的伞状松树林之间的对比，一种无法言语的对话被建立了。整个场地被浓密得无法穿越的树篱标识出来，形成了场所一种可见的围合。本项目极具个性，它把受军事考虑而形成的解决办法，与一种效果开放且受欢迎的轻松形式相结合。在场地内部，哈格安排了隐藏的路径，这唤起了人们对运用在西雅图联邦大厦中几何形解决方案的回忆（景观建筑师与弗雷德·巴塞蒂之间又一次主要的合作）。哈格再一次进行了艺术创新，如此谨慎地提供了一种人工解决方案，以至于在整体布置上完全看不出痕迹，这可是个不易实现的目标。

Richard Haag's one project in Europe, the project for the US Embassy in Lisbon involved the redevelopment and redefinition of existing spaces. The architect Fred Bassetti exploits the difference in level between Avenida das Forças Armadas and the slightly raised site of the old aristocratic villa to create a cascade of architectural volumes that house the various services facilities of the embassy and consulate. It is a simple and regular layout that combines the clean rational lines typical of Bassetti's designs with the security requirements laid down in State Department specifications. Within this enclosed and introverted ensemble, Richard Haag works to shift the point of view away from the built-up structures to the huge mass of Mediterranean pines that occupies the centre of the embassy grounds. Whilst adding to the privacy of the building behind – which it almost totally conceals from the street – this mass of trees also becomes a delightful device serving to establish communication between the internal and the external. An unscripted dialogue is set up thanks to the contrast between the traditional roof tiles and the dense green of the umbrella pines. The entire space is marked out by a thick, impenetrable hedge that forms the visible enclosure of the property. Highly institutional in character, the project combines solutions inspired by military considerations with a lightness of form whose effect is open and welcoming. Within the grounds, Haag lays out sheltered pathways that are reminiscent of the geometrical solutions adopted for the Federal Building in Seattle (the other major collaboration between the landscape architect and Fred Bassetti). Once again Haag puts his art to new use, providing working solutions that are so discreet as to be almost completely invisible within the overall layout as a whole – a goal which it is no easy thing to achieve.

U.S. EMBASSY LISBON PORTUGAL

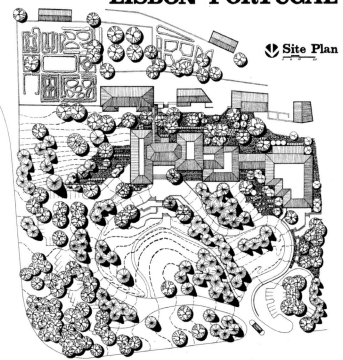

Site Plan

FredBassetti & Company Architects

RICHARD HAAG ASSOCIATES, INC.
landscape architects ····· site planners
701 northeast northlake way, Seattle, WAshington, 98105 (206) 634-1020

21.

维克托·施泰因布吕克公园

北纬：47°36′37.15″
西经：122°20′38.30″

地点：西雅图，华盛顿州，西大街 2000 号
时间：1982 年
状况：已建成
项目设计：维克托·施泰因布吕克
项目经理：理查德·哈格
设计团队：理查德·哈格事务所（佩吉·盖纳）
项目委托方：Seattle Parks Department
摄影师：Mary Randlett
艺术家：Marvin Oliver, Jim Bender (Totem Totem Poles); Roman E. Torres (Ferrobattuto Ironwork); Buster Simpson (Scultura rampicante Climbing Sculpture)
场地面积：0.3 公顷（0.8 英亩）

在西雅图与美丽的皮吉特湾之间并没有直接的联系，城市终止于一个陡峭的高差变化，这个峻峭的下降划分了城市街区和港区。道路本身也反映了这个地理上的分裂：在陡坡的下方，是紧邻海岸线的阿拉斯加路，而在陡坡的上方，则是第一大道，刚性的栅格划定了城市的布局。两条路之间的连接是派克广场，一个过渡地带现在坐落着很多"过时建筑"，通过台阶和电梯来联系海岸和城市。著名的派克市场是一个由木材和钢材为构架的市场，官方开业于 1907 年，这个欧洲风格的市场是绿宝石城的一个标志。然而，在 1971 年派克市场经历了拆迁危机，在大众对它的价值认知被诸如理查德·哈格和他的朋友维克托·施泰因布吕克（一位来自华盛顿大学的建筑师和资深学者）这样的艺术家们唤醒后才得以保留。两位艺术家一起设计了市场公园，在 1985 年施泰因布吕克去世之后，哈格接着致力于该项目。这个公共空间修建于一座多层停车场的楼顶上，具有一个深思熟虑的无明显特征的布局，既为市场购物设施创造了充分的空间，又成为了眺望奥林匹克山和海湾的观景楼。实际上，这里是相对地平线的交叉点。在儿童游乐区和供人们交流的座位区中间，是一个高低起伏的草坪，延伸到不同的水平面。装饰的两根图腾柱是一个象征着过去、现在和将来纽带的艺术品。作为实质性公众参与的成果，这是个简单到几乎保守的布局，设计师从中反映了当地商人的需求和本地市民的渴望。这个特征突出的地方，经历时间流逝而不改变，施泰因布吕克公园依旧是西雅图民众广泛喜爱的场所。

There is no direct relation between Seattle and the marvellous Puget Sound; the city ends in an abrupt change in elevation, a sheer drop that has always divided the urban fabric from its port. The roads themselves reflect this geographical rupture: below is the Alaskan Way, which follows the line of the coast, above is First Avenue, which defines the rigid grid of the urban layout. The connection between the two is provided by Pike Place, a sort of middle ground which is now the location of a number of 'through' buildings, with steps and lifts serving to link sea level and city level. Upon this framework of wood and metal, the famous Market developed; officially opened in 1907, this European-style market would become one of the symbols of the Emerald City. However, in 1971 Pike Place Market actually risked demolition and was only saved when general awareness of its value was re-awakened by such activists as Richard Haag and his friend Victor Steinbrueck, an architect and fellow lecturer at the University of Washington. Together these two would design Market Park, which was dedicated to Steinbrueck after his untimely death in 1985. This is a public space created on the roof of a multi-storey car park; a deliberately anonymous layout, it was created to be both a back-up/overflow space for the market's shopping facilities and a belvedere offering views over the Olympic Mountains and the bay. In effect, it is a point of contact between opposite horizons. In the midst of playing areas for children and seating designed for social interchange is an undulant lawn; extending over various shifts in level, this is adorned by two totem poles that serve as artistic symbols of the links between past, present and future. The fruit of a true process of public participation, this is a simple – almost timid – layout, in which the designers reflect the needs of the local traders and the desires of local citizens. An area of clear character that will stand unchanged over time, Steinbrueck Park is still a popular haunt for the people of Seattle.

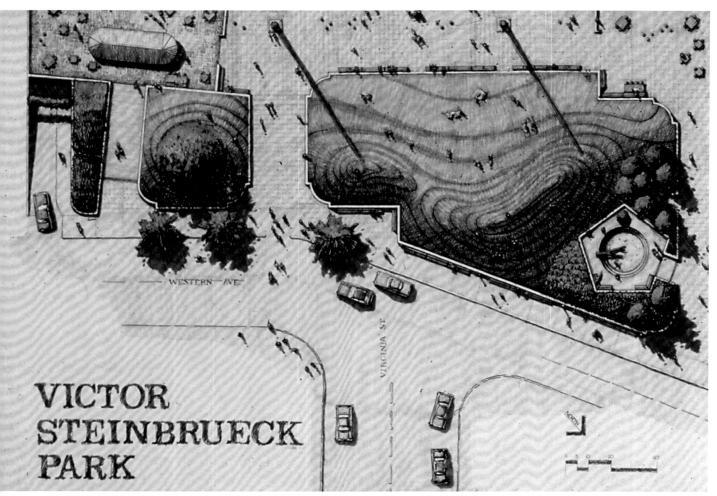

VIEW LOOKING WEST

VICTOR STEINBRUECK PARK

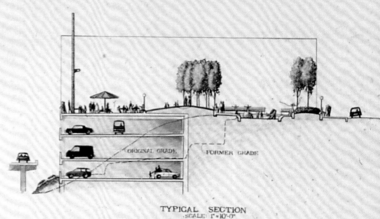

TYPICAL SECTION
SCALE: 1"=10'-0"

22. 狄龙住宅

地点：贝尔维尤，华盛顿州
时间：2004 年
状况：已建成
项目设计：苏利文／康纳德
项目经理：理查德·哈格
设计团队：理查德·哈格事务所
项目委托方：Eric + Holly Dillon
主要承建方：Signature Quality Landscape (John Kroetch)
摄影师：courtesy RHA, Luca M.F. Fabris

最近几年，理查德·哈格事务所为西雅图的因互联网经济繁荣和传统商业转型而形成的富裕阶层，广泛地设计庭园和房屋四周的空地。狄龙住宅位于沿着华盛顿湖东岸发展的城市贝尔维尤，这里不连续的住宅沿着海岸线延伸，属于精英住宅区。这个项目完工于 2004 年，委托方提供给理查德·哈格丰厚的预算。然而，设计师考虑到这是一个古典主义的建筑，因此选择了比较简约的方式来处理建筑周围的场地。通往建筑的通道由桦树的树叶墙限定，那些白色细长的树干，衬托在两片隆起延伸的草地中，草地包含在车道两边，由专门从中国进口的像月亮一样洁白的石头砌成矮墙。住宅的周围，许多小的开放空间形成于各种布局的构筑物之间，通过设计成为小庭园和院子；设计重新创造出一种既具有典型英国风味又有地中海气息的风格。设计者在场地的角落分别布置了水景园、玫瑰园和一个单纯的花园。每一个这样的空间都得到了很好的占用和使用，不仅仅是一种设想而已。在场地靠湖的一面，花园的斜坡陡然向下，场地布局由于诸如台阶下沉在草地里的这些特征而变得精彩。整个布局的主要特征是一棵巨大的本地松树。作为场所精神的标志，它超越尺度地直接竖立在住宅的对面。其他方面，项目还十分注重细节以赢得精致的设计，比如说，通过手工制作把石头挖空，当做花盆，里面栽植一些攀缘植物来装饰整个建筑。

In recent years Richard Haag Associates has worked extensively on the design of gardens and grounds for the those whose wealth is linked with the boom of the Net economy and the shift in the traditional industrial activities associated with Seattle. Dillon Residence is located at Bellevue, a city which has developed upon the eastern shore of Lake Washington and has become a prime residential area, with uninterrupted housing extending along the line of the coast. Completed in 2004, this project allowed Richard Haag to take advantage of a sizeable budget; however, working around decidedly classicist architecture, the designer opted for an approach that is almost minimalist. The access route to the house is framed by leafy walls of birch trees, whose slim white trunks stand out against the two swelling expanses of lawn contained at the sides of the drive by low walls in lunar-coloured stone specially imported from China. Around the house, the small open spaces created between the various sections of the structure become small gardens and courtyards, recreating something that has both a typically British flavour and a hint of the Mediterranean. One corner is dedicated to water, another to a rose garden, and yet another to a garden of simples. Each one of these spaces is to be occupied and used, not merely contemplated. On the lakeside of the property, the garden slopes downwards sharply, the lay of the land being highlighted by such features as the steps sunk within the grass of the lawn itself. The dominant feature in the entire layout, however, is the enormous local pine tree; a monument to the spirit of the place, this stands (almost out of scale) directly opposite the house itself. Elsewhere the project is predicated upon small details that are achievements of refined design – see, for example, the hand-worked stone which has been hollowed out to serve as a plant-pot for the creepers that adorn the building.

23. 蓝绿公园竞赛

北纬：33°40′23.56″
西经：117°43′47.00″

地点：奥兰治县，加利福尼亚州
时间：2005 年
状况：未建成
项目设计：美国 Ectypos 建筑工作室；意大利珀兹欧·博欧利工作室
项目经理：齐瑞尔·瑞维森
设计团队：理查德·哈格事务所（杰克·杰克森，设计者；利兹·韦斯特布鲁克，排版人，杰·路德，制图人）
项目委托方：Orange County Great Park Corporation
摄影师：courtesy RHA
场地面积：400 公顷（1000 英亩）

弃置地的再开发在理查德·哈格项目中是重要的组成部分，在他的杰出成就中拥有显著的影响。在 2005 年，理查德·哈格与 Ectypos 建筑工作室共同参加了奥兰治县废弃军事基地的再开发设计竞赛。竞赛使他能实验和更新自己在这个领域发展的设计语言，并进一步强化在这些项目中可持续性的地位与角色。在总体布局中，军事跑道将被保留，而方案富有想象力地将新的住宅区沿场地的西边布局展开，这里曾经是基地的后勤设施。整个场地中心是蓝绿公园的两枚绿色"长钉"，呈十字交叉。理查德·哈格富于想象的设计不仅仅是一个线性规划，而是一个可供社区各种活动使用的公园和绿地系统。整个设计在"格朗德山"达到顶点，作为场地的制高点，在这里日晷的指针或图腾柱标志着时间的流逝。使用来自军事基地拆除的碎石和挖掘格兰特湖与建筑基础得到的材料，重新塑造场地的形式，在很多的方面都会使人联想到阿兹特克金字塔。这里有从城市的到乡村的各种绿化空间。位于整个场地南部和北部的两个长型且几乎对称的花园，规划在城市肌理中，然后被细分为一系列的开放广场。这些直角的花园使场地的中心蕴含了更多的英式韵味。这里地形起伏，广袤的草地和林地绵延交替。公园的另一个关键特征是水景，水源由场地改造形成的水道和大片住宅区收集的雨水提供。格兰特湖的一边，方案设计了一个圆形剧场，向着原始的大地开敞，使其回复自然的韵律。然而，蓝绿公园可能只是一张蓝图，一个可能的乌托邦项目。

The redevelopment of disused areas has always played a major part in Richard Haag's work and has had a clear effect upon his major achievements. In 2005 the competition together with Studio Ectypos Architects for the redevelopment of a disused military base in Orange County (California) would enable him to experiment with and update the lexicon he had developed in this field, stressing even further the role that sustainability should play in such projects. The general layout of military runways was to be maintained, while the scheme envisaged new housing laid out along the western side of the site, which had once housed the base's logistics facilities. At the centre of that whole area was to stand the two green 'spikes' of Blue Green Park, forming a sort of cross. Rather than a linear layout, what Haag envisaged was a system of parks and green spaces which the community could use for a variety of activities; the whole would culminate in the Grand Knoll, the highest feature of the terrain, where a gnomon/totem marked out the passage of time. In some ways reminiscent of Aztec pyramids, the form of the remodelled terrain was to be achieved using the rubble from the demolition of the military base and the material excavated in digging the foundation for the Great Lake and housing complex. There are various types of green spaces here, from the urban to the rural. Lying to the north and south of the entire site, two long, almost symmetrical, gardens are laid out on an urban scale and subdivided into a series of open piazzas. At right angles to these was to be a central area of more English inspiration; here the terrain would rise and fail, with wide stretches of grass alternating with woodland. The other key feature of the park was to be water, provided both by the watercourses brought to light during work on the site and also by the rain water collected within the large housing estate. The Great Lake, on one side of which was to be laid out an amphitheatre, opened towards uncultivated ground restored to the rhythms of Nature. Blue Green Park would, however, remain on the drawing-board, a project for a possible utopia.

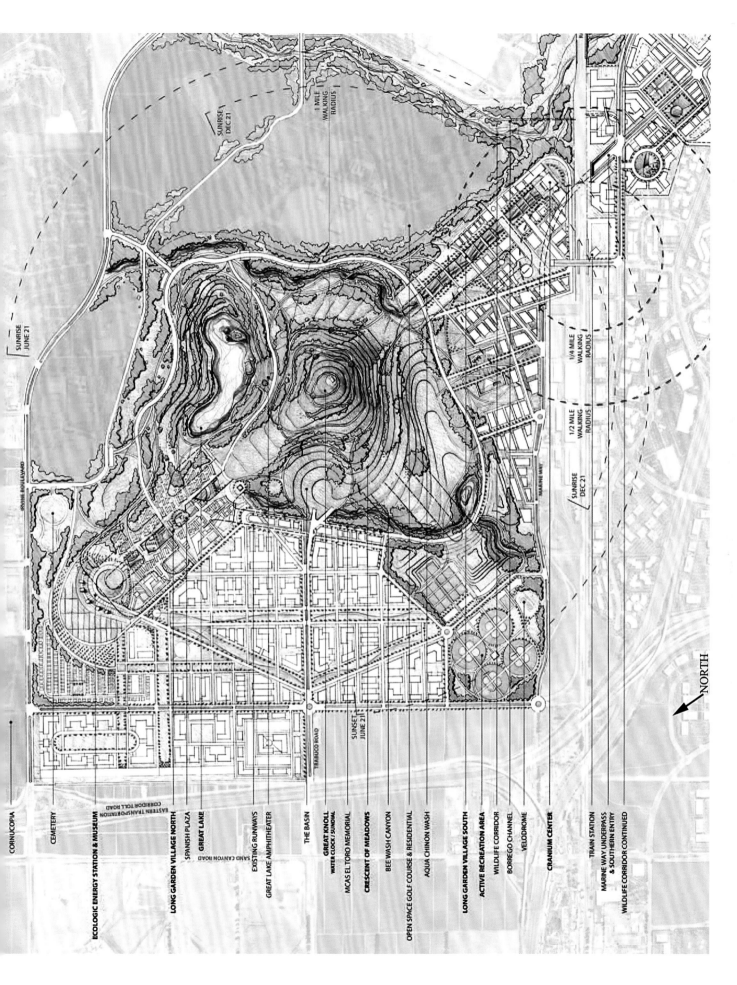

24.

库克斯住宅

地点：贝灵汉，华盛顿州
时间：1986～1990年
状况：已建成
项目设计：大卫·霍尔（亨利·克雷因伙伴公司）
项目经理：理查德·哈格
设计团队：理查德·哈格事务所
项目委托方：Wilbur + Dinah Kukes
主要承建方：Tom Weans (carpenteria masonry)
摄影师：courtesy RHA

为了体现与自然的直接接触，景观项目利用取自霍特科姆湖的石头，砌成库克斯住宅的牢固基础。这座木制小别墅被认为是建筑师大卫·霍尔评价最高的作品之一，他是亨利·克雷因伙伴公司（位于弗农山庄，华盛顿州）的成员。受流行的北美湖边度假别墅的启发，库克斯住宅被用于常年居住，它较大的体量被清晰的形态掩饰。哈格设计了由水泥薄片构成的船坞，看起来像浮在水面上的一样，赢得了一种看似简单却十分令人震撼的效果。这个非自然的产物看起来变成了自然的一部分，在混凝土（人造石头）和大量露出水面被植物覆盖的棕色岩石之间建立了一种对话。由岩石和石块组成的小山用于界定入口的空间，包围住宅的整个场地的布局，呼应着1962年斯塔米诊所设计中使用过的形式。理查德·哈格在这个项目中使用了大量而厚重的岩石，为以绿色针叶树作为背景的微观世界，构成一个缓慢而柔和的界定。整个项目让石头叙述自己，形成安静而放松的氛围。带有更新的新鲜感，石头再次告诉了我们，方案受到日本园林的启发。同时，这种根源也反映在了船坞的设计中，船坞引发了人们的无限联想，小船坞看起来像通往另一个世界的门口。清晰的形式在断言，哈格在项目中几乎没有运用突出的特征，而是让其本身受自然引导。人类的脆弱强调了人存在于与环境的平等相处之中，而非统治它。

In direct contact with Nature, this landscape project exploits the rock which, emerging from the waters of Lake Whatcom, becomes the solid base for the Kukes Residence, a wooden 'cottage' that is still one of the most highly-regarded designs by the architect David Hall, a partner in the Henry Klein Partnership (Mount Vernon, Washington). Created to be inhabited all the year round, the house is inspired by the holiday cabins that are so common on the North American lakes, its much greater size being disguised by the articulation of the volumes. Designed by Haag, the dock is made up of thin sheets of cement that seem to fly out over the water, achieving an effect that is powerful in its simplicity. The non-natural seems to become part of nature, establishing a dialogue between the concrete (man-made rock) and the brown mass of rock that emerges from the water to be overwhelmed by vegetation. The layout of the area around the house echoes forms that had already been used in the design for the Stamey Clinic (1962), with small mountains of rocks and stones serving to define the access points. Richard Haag here plays with the mass and weight of rock to generate a slow and gentle development that defines a parallel microcosm, with the green of the conifers serving as a backdrop. Calm and relaxed, the whole project allows the stone to speak for itself. With renewed freshness, this material once again speaks to us of the schema derived from Japanese gardens. And that source is also reflected in the dock itself, which becomes an imaginary bridge towards the infinite, with the small boathouse seeming a doorway to another world. Predicated upon clarity of form, these few features are exploited by Haag in a project that allows itself to be guided by Nature herself; the fragility of the human highlights that mankind exists in parity with – not dominance over – his environment.

25.

萨默维尔住宅

地点：梅迪纳，华盛顿州
时间：1987 ~ 1990 年
状况：已建成
项目设计：罗伯特·斯莫尔
项目经理：理查德·哈格
设计团队：理查德·哈格事务所（罗伯特·弗雷）
项目委托方：Mimi + Vinton Sommerville
主要承建方：Tom Paulsen
摄影师：Luca M.F. Fabris
艺术家：Charles Greening (ruscello stream corridor)

理查德·哈格为萨默维尔住宅所设计的花园是一个惊喜。位于梅迪纳的该住宅，伫立在埃弗格林地区的良好位置上，通过西面的水面可以看到西雅图的天际线。这栋住宅经过一段弯曲的道路到达，几乎隐藏在其他的住宅之中。在哈格的巧手下，地势的连续下降成为了吸引人的雕塑作品。路面由砖铺砌，这个独特的设计意图提供一种齿条和齿轮的效果，因为在冰雪时候，光滑的路面会产生危险的打滑。倾斜的砖块和它们投下的阴影形成双轨迹导致岔路：一个方向带你向上到房子那里，另外一个通向草坪，在华盛顿湖的边缘，变成狭窄的鹅卵石沙滩。这个奢华的房子有着宽大的窗户，看起来几乎像个温室。室内外混合在一起，花园变成了生活空间的一个延伸。简单的方形瓷砖布置在草坪上形成一个开放空间，然后随着一步步远离这座房子而逐渐变得稀疏。再次，哈格从日本庭园中引入特征，为了强化建筑物的组织，经常完全摒弃任何巴洛克风格的暗示。一些高大的落叶树沿着房子的一边种植，为夏天的午后提供遮阴。在南面，地面被延伸向湖边的树篱围合，长成一个灌木的杂树林。林中有个小池塘，池塘里的流水顺着小溪，水源来自位于房子后面的雕塑喷泉。对于哈格而言，这条小溪蕴含着某种寓意。它的声音和外表表达了一种生命力的感觉，融化在微观世界中，然后与更广范围的水体融为一个整体。

The garden that Richard Haag created for the Somerville Residence is a surprise. Located at Medina, the house stands in a fine position on Evergreen Point, which affords a view of the Seattle skyline over the water to the west. The house is reached by going down a curving drive, which is almost hidden between the other properties. In Haag's hands, this unbroken descent becomes an attractive work of sculpture. Made in brick, its unusual design is intended to provide a sort of rack-and-pinion effect on the days when ice and snow would make a smooth surface dangerously slippery. The sloping bricks and the shadows they cast form a double track which leads to a fork; one direction then takes you up to the house proper, the other towards the lawn which, at the edge of Lake Washington, becomes a narrow beach of pebbles. This luxury house has such vast windows it almost looks like a greenhouse. The interior and exterior blend together, with the garden becoming an extension of the living space. Simple square tiles laid on the lawn form an open space, then gradually thin out as one moves further and further away from the house. Once again Haag introduces a feature from the Japanese garden in order to lightened the 'weave' of a fabric which (as always) totally eschews any hint of baroque emphasis. Some tall deciduous trees planted alongside the house serve to shelter it from the sun's rays in the summer afternoons. To the south, the grounds are bound by a hedge which, as it extends towards the lake, becomes a copse of shrubbery. This latter contains a small pond into which flows the stream carrying water from the fountain-sculpture located behind the house. For Haag, this small watercourse has a sort of allegorical import. Its sound and appearance express a sense of life itself, of a small microcosm which dissolves and then forms a single whole with the larger expanse of water.

26.

泰特·弗里斯住宅

地点：西雅图，华盛顿州
时间：2005 年
状况：已建成
项目经理：理查德·哈格
设计团队：理查德·哈格事务所（理查德·格瑞沃斯）
项目委托方：Richard Tait + Karen Fries
主要承建方：Signature Quality Landscape (John Kroetch)
摄影师：courtesy RHA, Luca M.F. Fabris

在华盛顿湖西侧的一个海角项目，理查德·哈格创造了一个光线充足的空间，反映了业主活跃的特点——成功的儿童游戏发明家（还不仅仅是儿童）。地产的路旁界限被树篱标识，面向单车道小路扩大而形成了一片前院。古典的外观和相当严格的对称使房屋本身就变成了一个特大号圆形剧场的舞台，由两级台阶的下沉组成，被大量的针叶树的树荫所覆盖。这里，景观设计师种植了鳞茎的和开花的植物，以保证草坪在一年四季都点缀着色彩。在灌木中的雏鸟是孩子们的玩伴。同时，在高差最后变化的末尾是一大片草坪，草坪结束于中间布置着青铜雕塑的砖铺地。开放的区域有水柱和喷泉，业主可以随意将其打开，将停车区域转变为一个夏日娱乐的场所。整个项目避免了强势和说教，取而代之的是，设计叙述着单纯的愉悦，基于各种功能需求的混合而创造出一个有趣而实用的空间。这是为年轻人而建的花园，能够让人无所顾忌地享受。实际上，这里没有小径和边界的踪迹；所有的空间看起来相互渗透，与来自于数学理论的规则相协调。一些落叶树用来为房子遮蔽过多的阳光，也创造了绿色的帘幕，更增加了远景深度感。在两侧，房屋俯瞰着湖泊，景观由迷迭香和山茶花植物构成，在远处融入森林，形成一个连续的视野。住宅坐落于西雅图，但却享受到了地中海的温暖和芬芳。整个设计是一个让人着迷且惊奇的透彻简述。

On a promontory projecting from the western side of Lake Washington, Richard Haag has created a light-filled space that reflects the effervescent character of the owners, the successful inventors for games for children (and not only children). The roadside boundary of the property is marked by a hedge, which opens onto a single-lane driveway that widens into a forecourt. Classical in appearance and rather stiffly symmetrical, the house itself becomes the proscenium for a outsize amphitheatre made-up of two stepped depressions shaded by enormous conifers; here, the landscape architect has planted bulbs and flowering plants which guarantee that the lawns are dotted with colour throughout the various seasons of the year. Nestling within the shrubs are play structures for children. And at the foot of the last change in level is a large lawn that ends in a brick surface whose centre is occupied by a sculpture in bronze. The open area has water jets and fountains which the owners can turn on at will, transforming the parking area into an entertaining summer feature. The entire project avoids the emphatic and didactic; instead, the design is predicated upon pure delight, upon the combination of functional needs with a playful use of space. It is a garden for the young, to be enjoyed without any preconceived ideas. In effect, there are no traced out pathways or boundaries; all the spaces seem to interpenetrate in accordance with rules derived from set theory mathematics. Some deciduous trees serve to shield the house from excessive sunlight and also to create a curtain of greenery that adds a greater sense of perspective depth. To the sides, the houses overlooks the lake; these views are framed by rosemary and camellia plants which, in the distance, blend into the forest, forming an unbroken visual field. The house is in Seattle, but one enjoys the warmth and the perfumes of the Mediterranean. The whole design is a transparent narrative by which one is both enthralled and surprised.

理查德·哈格自传
An Auto-biography
by Richard Haag

理查德·哈格于1923年10月23日出生在肯塔基州的路易斯维尔市。他的父亲是鲁迪·哈格，母亲是路特拉·奥因斯·哈格。鲁迪·哈格是第二代德国人，路特拉的家庭是第三代的德国人和第四代的英国人。理查德是六个孩子中最大的，他还有两个弟弟和三个妹妹。

理查德的父亲鲁迪·哈格是个自学成才的园艺师和园丁，并且在1920年建立R·L·哈格苗圃，在这里工作了50年后，于1985年卖掉了这个苗圃。

哈格从童年时期开始就十分喜欢植物和户外活动，并且经常独自在苗圃、附近的树林和小溪边漫步。理查德的植物学知识十分丰富，他四岁的时候，一篇有关他的植物嫁接技术的文章在当地报纸上发表了。

1942年理查德入伍美国空军，并且被培训为雷达工程师，在中国、缅甸及印度的战区供职直到1945年退伍。

退伍军人权益法案使理查德在伊利诺伊大学厄巴纳－香槟分校开始正式的景观建筑学学习。在那里他师从两位最有影响的景观建筑学教授——斯坦恩雷·怀特和佐佐木英夫。

理查德不得不在两位老师截然不同的思维方式中找到平衡点——怀特的设计方法非常浪漫而且灵感丰富，相比之下佐佐木的设计方法则更注重理性和科学。理查德和两位老师保持了毕生的友谊。在1948年夏天，理查德和斯坦恩雷·怀特在新英格兰地区旅行，参观博物馆、植物园和许多著名的花园。第二年的夏天，理查德和佐佐木英夫穿越整个国家到加利福尼亚旅行。两次旅行的交通工

Richard Haag was born in Louisville, Kentucky October 23, 1923. His father was Rudy Haag and his mother was Luthera Owings Haag. Rudy Haag was a second generation German and Luthera's family was third/fourth generation German/English. Richard was the eldest of 6 children – 2 brothers and 3 sisters.

Richard's father, Rudy Haag was a self-taught horticulturist/nurseryman and started R.L. Haag Nursery in 1920, worked the nursery for 50 years, the nursery was sold in 1985.

From childhood Richard had a love of plants and the out-of-doors and was very much left on his own to roam around the nursery and nearby woods and streams. Richard's knowledge of plants was exceptional and when he was four years of age an article about his plant grafting techniques was published in the local newspaper.

Richard enlisted in the US Air Force in 1942. Trained as a radar engineer he served in the China-Burma-India Theater until discharged in 1945.

The GI Bill of Rights enabled Richard to begin formal studies in Landscape Architecture at the University of Illinois, Champaign –Urbana where he studied under two of the most influential Professors of Landscape Architecture, Stanley White and Hideo Sasaki.

Richard had to find a way to balance their different ideologies – White's very romantic and inspirational contrasted with Sasaki's rational and scientific approach to design. Richard developed life-long friendships with both teachers. In the summer of 1948 Richard and Stan White toured New England, visited museums, arboretums and famous gardens. The following summer Richard traveled across country with Hideo Sasaki to California. Mode of transportation for both trips was Haag's open military jeep.

Richard transferred to the University of California at Berkeley to study under Professor Robert Royston to get into more progressive/modern action and gradua-

具都是哈格开敞的军事吉普车。

理查德转学到了加利福尼亚大学的伯克利分校，跟随罗伯特·罗伊斯顿教授学习更为先进更为现代化的方法，并且在1949年取得景观建筑学的学士学位。毕业后的夏天理查德来到密歇根州跟随佐佐木英夫实习。

1950年，理查德进入哈佛大学设计研究院的研究生工作室，并且致力于与建筑师、城市规划师和同行的景观建筑师团队合作设计。理查德在1952年得到了他的景观建筑学硕士学位。在哈佛的夏天，理查德跟随景观建筑师丹·基利在弗朗科尼亚的新罕布什尔州工作室工作。

毕业之后理查德为阿巴拉契亚（美国肯塔基州和弗吉尼亚州）的美国煤矿工人医院，开始了长达两年时间的场所规划咨询和景观建筑设计。

在1954～1955年理查德获得了富尔布莱特基金的奖学金而赴日本京都大学学习。走遍日本的学习和旅行经历，对理查德的职业生涯和个人生活方面都有终身的深远影响。

其后，理查德返回旧金山，与在伯克利的同学理查德·维格罗拉、杜·卡特和萨图茹·尼西塔一起为劳尔·哈普林工作。

通过考试，理查德被授予了加利福尼亚州景观建筑的职业资格证书，并于1957年在旧金山成立了理查德·哈格事务所。

经过了40个项目，1958年理查德接受了在西雅图的华盛顿大学建筑与城市规划学院的教职。他在学院中建立了景观建筑学系。该系六年后得到了美国景观建筑师协会的一致认可。

ted with Bachelor of Science Landscape Architecture [BLA] in 1949. The summer after graduation Richard apprenticed with Hideo Sasaki in Michigan.

In 1950 Richard entered the Master's studio at Harvard Graduate School of Design and concentrated on collaborative design with teams of architects, urban planners and fellow landscape architects. Richard received his Master of Landscape Architecture [MLA] in 1952. During the summer months while at Harvard, Richard worked in the Franconia, New Hampshire office of Landscape Architect Dan Kiley.

Following graduation Richard began two years of consultation site planning and landscape architecture designs for the United Mine Workers' hospitals in Appalachia [US States Kentucky and Virginia].

In 1954-55 Richard was awarded a Fulbright scholarship to study at Kyoto University, Japan. Studies and travels throughout Japan had a profound lifelong influence in Richard's professional and personal life.

Richard returned to San Francisco and worked for Larry Halprin with former Berkeley classmates, Richard Vignola, Don Carter, and Satoru Nishita.

After examination, Richard was granted the California Landscape Architectural Professional License and in 1957 incorporated Richard Haag Associates in San Francisco.

Forty projects later, in 1958, Richard accepted a position in the College of Architecture and Urban Planning at the University of Washington in Seattle. His role was to establish a Department of Landscape Architecture in the College. The Department received full accreditation by the American Society of Landscape Architects six years later.

Richard moved Richard Haag Associates Inc. to Seattle, Washington and in 1962 Richard purchased 40 acres and began Trees Nursery to grow specimen trees.

理查德将理查德·哈格事务所（RHA）搬到华盛顿州的西雅图市，并在1962年购买了40英亩的土地，开设了一个培育样本树种的苗圃。

西雅图在20世纪60年代早期蓬勃发展，而理查德·哈格事务所是当地唯一的专业景观建筑公司。早期的委托方是来自西北部现代建筑运动的领袖们——建筑师弗里德·巴塞蒂、鲍·希瑞和鲍·柯克。项目包括大学校园的国家级优胜奖、私人和公共机构在1962年对举办了世界博览会后的西雅图中心区的改建，使其成为了西雅图城市主要的开放空间和文化中心。

理查德·哈格事务所第一批雇员，招聘的是理查德在建筑系场地规划课程中教授的学生。格兰特·琼斯、波布·汉娜、劳瑞·奥林和弗兰克·杰姆斯都是一些早期的学生和雇员。他们后来都成为了这个专业领域里的执业者和领袖。汉娜、奥林和杰姆斯既从事景观建筑的实践也从事教学。

理查德另一个巨大的推进是在公园规划与设计领域。他赋予自然环境中的自然材料、土地的形式以及孩子们相互玩耍的一般设计构造以显著的特征，并在公园设计中予以表达新的设计语言——循环利用遗迹、寻找震撼处、表达孩子们为何游戏的精炼理论（每个孩子都应该有机会挖掘、攀爬、隐藏、寻找和在水中嬉戏）。

该进程以煤气厂公园（1970-1975年）和布洛德尔保护地（1969/1980-1986年）这两个项目达到顶点。在这个时期和90年代初期的其他著名项目，还有美国驻葡萄牙里斯

Seattle was booming in the early 1960s and Richard Haag Associates, Inc. [RHA] was one of the only professional Landscape Architectural offices on the scene. Early commissions came from leaders of Modern Northwest Architecture movement, Architects Fred Bassetti, Paul Thiry, and Paul Kirk. Projects included national award winning University campuses, private and public institutions, and the conversion of the 1962 World's Fair into Seattle Center, the City's premier open space and cultural center.

RHA's first employees were students recruited from Richard's site planning classes taught to architecture students. Grant Jones, Bob Hanna, Laurie Olin and Frank James were some of the early student/employees. All went on to become international practitioners and leaders in the profession. Hanna, Olin and James taught as well as practiced Landscape Architecture.

Richard's next big push came for public park planning and design He featured natural materials in natural settings, earth forms, customed designed structures for interactive children's play, and gave expression to a new vocabulary in park design – recycled ruins, thrill seeking, giving expression to the recapitulation theory of why children play. [Every child should have opportunities to dig, climb, hide, seek, and play in water.]

This progression culminated with Gas Works Park [1970-75] and Bloedel Reserve [1969/1980-86]. Other notable projects at this time and into the early 1990s were the U.S. Embassy and Consulate, Lisbon, Portugal; Jordan Park [Everett Marina Park]; Victor Steinbrueck Park; Summerville, Kukes, Stern, Bluhm Residences; Merrill Court Townhomes; Ameriflora – international garden exposition; Berkeley North Waterfront Park, and University of California Berkeley Memorial Glade.

More recently Richard and his teams were finalists [and favored by public vote] for two major competitions: Blue Green Orange County Park, California and 2008 Beijing Summer Olympics. Dillon and Tait-Fries Residences, and the Frye Art Museum were completed

本的大使馆和领事馆；乔丹公园（埃弗里特码头公园）；维克特·施泰因布吕克公园；萨默维尔、库克斯、史特恩、布鲁姆住宅；梅里尔庭院式联排住宅；美洲植物与国际园艺博览会；伯克利北部滨水公园和加利福尼亚大学伯克利分校林间纪念广场。

最近一段时期，理查德和他的团队入围了（受公众喜爱投票胜出）两个重要的设计竞赛：加利福尼亚州奥尔治县蓝绿公园和北京 2008 年夏季奥运会。狄龙和泰特·弗里斯住宅以及弗里艺术博物馆都完成于 21 世纪初期。目前，理查德正忙于还没有向公众公布的内部项目。然而，他在成都的作品，中国 XWHO 设计允许他毫无拘束地发挥自己的设计技巧。

理查德写作、演讲并担任国内与国际设计竞赛评委。理查德也因他批判的实践主义而闻名——拯救了派克广场的市场，致力于生态和环境问题以及社会公正。

in the early 2000s. Currently, Richard has been busy with confidential projects not available for publication. However, his work in Chengdu, China with XWHO Design allowed him to give free reign to his design skills.

Richard writes, lectures, and serves as competition design juror nationally and internationally. Richard is well noted for his civic activism – saving the Pike Place Market, ecologic and environmental issues and social justice.

附录
Appendix

GAS WORKS PARK

Bibliografia (parziale) *Bibliography (partial)*

-, "A Park From Industrial Dinosaurs." *Compressed Air Magazine*, July 1988: 18-23.

-, "A Post Revolutionary Park is on its Way." *Archmuse* 11 [U of Washington, Dept. of Architecture] 8 Feb. 1972.

-, "A Tale of Three Cities." *Atlantic Monthly*, Apr. 1976: 65-6.

-, "Canoeing Close to Home – Around Seattle & Portland." *Sunset Magazine*, Apr. 1976.

-, "Gas Works Park." *Print Casebooks*, 4th ed., 1980-81.

-, "Gas Works Park." *Sunset Magazine*, Aug. 1977: 52-3.

-, "Gas Works Park: Recycling Giant Machinery into a Park." *Aesthetics in Transportation*. Washington D.C.: USDOT, 1980. 22.

-, "Il Parco del Gas." *Abitare*, Oct. 1984: 74-7.

-, "Machine-Scaping?" *L'Architettura*, Apr. 1978: 236-7.

-, "Seattle Captures Unique Multiple Honor of HUD." *Seattle Business Journal*, 16 June 1980.

-, "Seattle: City Life at its Best." *Sports Illustrated*, July 1982: 60 + 68.

-, "Sundial Dedication." *Seattle Arts*, Sept. 1978.

-, "The 1981 ASLA Professional Awards." *Landscape & Turf*, Nov./Dec. 1981: 20.

-, "The 1981 ASLA Professional Awards: "Gas Works Park – President's Award of Excellence." and "Profile: Rich Haag." *Landscape Architecture*, Sept. 1981: 544. 546, 548; 594-7.

-, "The Greening of the Gas Works." *Seattle Business*, 9 Jan. 1975.

-, "Usine à Gaz Reconvertie, Seattle, Washington." *l'Architecture d'Aujourdhui* 204, Sept. 1979: 74-5.

-, "Was einmal ein Gaswerk war…". *Baumeister*, Aug. 1979: 761.

-, *Urban Open Spaces*. Exhibit & Publ., Cooper-Hewitt Museum of Design. New York, 1979.

Advisory Council on Historic Preservation. "Industrial Dinosaurs." *Report to the President & the Congress of the United States*. 1984. Washington, D.C.: ACHP 27-7.

Bush, James. "Gas Works cleanup plans called excessive, costly." *The North Seattle Press*. 12/13-26/89: 1-2.

Byrnolson, Grace. "Gasworks (ugh!) Reborn as a City Park." *Smithsonian Magazine*, Nov. 1977: 117-120.

Campbell, Craig. "Seattle's Gas Plant Park." *Landscape Architecture*, July 1973.

Carpenter, Edward K. "Gas Works Park." *37 Design and Environment Projects*. Washington D.C.: R.C. Publications, 1976: 44-5.

Casey, Charlotte. "The Gas Works." *Washington Purchaser*, Apr. 1973.

Chapman, Bruce K. "The Gas Works: A New Look at an Old Eyesore." *Northwest Today* in *Seattle P-I*, 14 May 1972.

Chasen, Daniel Jack. "The Gasworks Revisited." *Pacific Northwest*, Jan./Feb. 1981: 8-11.

Collins, Alf. "Park Art or Cluttered Confusion, or …" *Puget Soundings*, Apr. 1970.

Cortesi, Isotta. "Gas Works Park." *Area*, July/Aug 1999: 78-87.

DeWolf, Jeff. "Gas Works Park Cleanup Cost Estimated in Millions." *The Seattle Press*, Jan. 15-28, 1997:1, 12.

DeWolf, Jeff. "Gas Works Park Cleanup Aired." *The Seattle Press*, Jan. 29-Feb. 13, 1997:7.

Dorpat, Paul. "Gas Works, a Garden of Metal." *orthwest Living* in *Seattle Times*, 12 June 1983.

Dorpat, Paul. "Stone Henge in Seattle". Seattle Times. Nov 26, 2006: 54.

Ehlers, Mariele. "Gasworks Park, Seattle." *Garten & Landschaft* (Journal for Landscape Architecture and Landscape Planning) [Germany] Aug. 1985:37.

Eller, Joe. "Playgrounds from Grimy Gasworks." *Christian Science Monitor*, 8 Aug. 1979.

Engstrom, Karen. "Lake Union Park: A Back-to-Nature Site." *Pictorial* in *Seattle Times*, 16 Dec. 1973.

Fabris, Luca Maria Francesco. *Il Verde Postindustriale*. Napoli: Liguori, 1999: 21-25.

Fels, Donald and Patricia Tusa. "Water at Gas Works can clean itself." *The Seattle P-I*. Aug 1, 2002.

Fels, Patricia Tusa (Friends of Gas Works Park). Application for Historic Landmark Status. Seattle Landmarks Preservation Board. 2002. Approved Ordinance #121043: 12/2002.

Fels, Patricia Tusa. "Gas Works at Work". *Arcade*. 19.4 Spring 2001.

Frankel, Felice, photographer and text by Jory Johnson. "Gas Works Park, Seattle, Washington." *Modern Landscape Architecture: Redefining the Garden*. NY: Abbeville, 1991. 196-207.

Friends of Gas Works Park. "Mission Statement" for 501-3C application. 1997 (approved).

Gantenbein, Douglas. "Design – West Coast Cities: Seattle." *The Columbian* [Vancouver, WA] 30 Oct. 1980.

Garten, Cliff. "Gas Works Park/Terrain Vague Hypothesis." Thesis, Harvard Graduate School of Design. 1998.

Gartler, Marion & Marcella Benditt. *The Talk of the Town*. (Photo Reader, Phoenix Reading Series) NY: Hall, 1980.

Getzels, Judith N. "Other Public facilities: Unexpected Opportunities." *Recycling Public Building report No. 319*. Planning Advisory Service, Am. Soc. of Planning Officials (Washington D.C.), Aug. 1976: 19-20.

Goldberger, Paul. "Gas Works in Centerpiece of Seattle Parks." *The New York Times*, 30 Aug. 1975.

Goldberger, Paul. "Ten Buildings With a Style of Their Own." *Portfolio*, June/July 1979:36-7.

Gunn, Clare A. "Vacationscape". 3rd Edition 1997: 133-134.

Haag, Richard. "Approved Myrtle Edwards Park Reviews List.", Jan. 1972.

Haag, Richard. "Gas Works Park – Phase 1 Construction." 1975.

Haag, Richard. "Influence and Power in Public Landscapes: The Gas Works Park Example." *GSD News* (Harvard University Graduate School of Design) Fall 1993: 26-7.

Haag, Richard. "It Was a Gas." *Outreach Magazine: the Urban Landscape* . Spring 1982.

Haag, Richard. "Memorandum re: Gas Works Park." 15 Apr. 1976.

Haag, Richard. "Seattle, Washington – Gas Works Park Fact Sheet." Sept. 1975.

Haag, Richard. "The Greening of an Ecologic Disaster – Gas Works Park, Seattle, Washington." (unpublished) Feb. 1993.

Haddad, Laura. Book Review: "Richard Haag: Bloedel Reserve and Gas Works Park". Arcade Journal for Architecture and Design in the Northwest. Seattle. Spring 1998.

Hargreaves, George. "Post Modernism Looks Beyond Itself." *Landscape Architecture*, July 1983: 64-5.

Harney, Andy Leon. "Refining a Refinery into a Park." *Trends*. Washington, D.C. Park Practice Program, Division of Federal and State Liaison, National Park Service, Jan./Feb./Mar. 1977: 29-30.

Hawkins, Robert. "Seattle, the Quality of Life." *Northwest Orient Passages*, Sept. 1974.

Helphand, Kenneth. "Generalscapes." *Landscape Architecture*, Nov. 1999: 83.

Hester, Randolf T., Jr. "Process Can Be Style." *Landscape Architecture*, May/June 1983.

Hester, Randy. "Labors of Love in the Public Landscape." *Places – Quarterly Journal of Environmental Design*. Vol. 1, No. 1, Fall 1983: 18 + 20-4 + 26-7.

Karol, John, filmmaker. *Working Places*. Society for Industrial Archaeology, 1975. (Available: National Trust for Historic Preservation, 1785 Massachusetts Avenue NW, Washington D.C. 20036).

Kidney, Walter C. "Gas Works Park." *Working Places: the Adaptive Use of Industrial Buildings*. Washington D.C.: Society for Industrial Archaeology, 1976: 135-7.

Kreisman, Lawrence. "Gas Works Park." *Made to Last*. Seattle: University of Washington Press, 1999: 74.

Macey, Daniel, Editor. "Playing Games with Park Place." *Gas Daily's NG Magazine*, Dec. 1995/Jan. 1996: 7.

Maryman, Brice and Scott Melbourne. "Gas Works Park." *Urban Beaches*. 2006.

McCullagh, James C., ed. *Ways to Play: Recreation Alternatives*. Emmanus, PA: Rodale P, 1978: 45-59.

Metzger, Jeane. "Old Gas Works to Become Unique Park." *Everett Herald-Western Sun Panorama*, 15 Jan. 1972.

Middleton, Michael. "Gas Works Park." *Man Made the Town*. London: Bodley Head, 1987: 211-3.

Miller, Lynn. "Ecology." *Landscape Architecture*. Nov. 1999: 62.

Okada, Atsuo. "Gas Works Park." *Nikkei Construction* (Nikkei Business Publications, Inc., Japan) July/Aug. 1994: 28-30, 41-2.

Okada, Masaaki. "Pioneer of Reusing Technoscape – Gas Works Park in Seattle." Special Issue on Industrial Landscapes. Japan Society of Civil Engineering, January 2009: 10-19.

Okada, Masaaki. "Technoscape: Theory of Integration and Estrangement." Book. Kajima Publishing, 2003.

Okada, Masaaki. "Techno-scape." Masters Thesis. University of Washington, College of Architecture and Urban Planning. 1994.

Oneto, Gilberto. "Archeologia Industriale nel verde" *Tuttoville*. 78, 1983.

Park, Chan-yong. "Gas Works Park in Seattle, U.S.A." *Korean Landscape Architecture*, July 1992: 56-61.

Pirzio-Biroli, Lucia. "Working Landscapes." *Arcade*. Winter 2004: 29.

Read, Kenneth E. "The Ghostly Gas Works." *Seattle Magazine*, Nov. 1969: 42-5.

Redstone, Louis G. *Public Art – New Directions*, Mc-Graw-Hill Book Co., New York, 1981: 41.

Reed, Peter. "Beyond Before and After: Designing a Contemporary Landscape." *Groundswell: Constructing the Contemporary Landscape*. New York: Museum of Modern Art, 2005: 14-32.

Reejhsinghani, Anju. "Gas Works Park Clean-up 29 Studies and Counting." *The Stranger*. Feb. 6, 1997: 8-09.

Richard, Michael. *Seattle's Gas Works Park: The History, The Designer, The Plant, The Park, Map, & Tour*. Seattle: Tilikum Place Printers, 1983.

Roberts, Paul. "Seattle Parks Under Fire: Landmarks Revisited." *Landscape Architecture*, Mar. 1993: 49-51.

Robertson, Hugh. "The Gas Works: How to Make a Park out of Recycled Jumbo Junk." *View Northwest* Sept. 1976: 38-40.

Rozdilsky, John. "The Landscape Art of Richard Haag: Roots and Intentions." Masters Thesis, U of Washington, Dept. of LA, Dec. 1991: 78-91.

Russell, James. "Landscape Urbanism." *Architectural Record*, Aug. 2001: 72.

Saunders, William S., editor. *Richard Haag: Bloedel Reserve and Gas Works Park*. NY: Princeton Architectural Press, 1998.

Scigliano, Eric. "Shaping the City." *The Seattle Times Magazine*, Nov. 10, 2002: 14-5.

Simo, Melanie. *100 Years of Landscape Architecture: Some Patterns of a Century*. Washington, D.C.: ASLA Press, 1999: 206-207, 222-223.

Spilk, et al. "Toward a New Way of Thinking about Urban Vacant Land." *Urban Vacant Land*. (The Pennsylvania Horticultural Society), Sept. 1995: 7, 41-3.

Steiner, Frederik. "L'architettura Americana del paesaggio: oltre i confini dei continenti." *Rassegna di Architettura e Urbanistica*. Dicembre, 1998: 53-8.

Steinitz, Carl. "Influences on Design Processes." *Defensible Processes for Regional Landscape Design*. Landscape Architecture Technical Information Series. Sep. 1979: 11+ 13.

Swift, Barbara & Jestena Boughton. "Richard Haag: Excerpts from an Interview." *Arcade* vol. III, no. 3 Aug./Sept. 1983: 3 + 9-10.

Thayer, Robert L., Jr. "Visual Ecology: Revitalizing the Aesthetics of Landscape Architecture." *Landscape* Winter 1976: 43.

Walker, Theodore. "Gas Works Park." *Designs for Parks & Recreation Spaces*. Mesa, AZ: PDA, 1987: 11.

Weems, Sally. "Gas Works Park." *Landscape Australia*, 1 Feb. 1980: 23-29.

Weilacher, Uto. "Syntax of Landscape", Birkhauser (Basel), 2008: 107-109.

West, Karen. "Dead Gas Plant Looks Toward Life as a Park." *Outlook* [Greenwood-Ballard Ed.]. Part 1: 13 Oct. 1971; Part 2: 20 Oct. 1971.

Weston, Richard. "Seattle Sculpture." *Building Design*, 20 Feb. 1987.

Willie, Deborah. "Public Art in Seattle." *Landscape Design*. No. 183, Sept. 1989: 43.

Woodbridge, Sally & R. Montgomery. *A Guide to Architecture in Washington State*. Seattle: U of Washington P, 1980: 59 + 206-207.

Woodbridge, Sally B. "Industrial Metamorphosis." *Historic Preservation*, Apr./June 1978: 37-41.

Woodbridge, Sally B. "It was a Real Gas." *Progressive Architecture*, Nov. 1978: 96-9.

Zoretich, Frank. "At Gas Works Park, Even the Grown-Ups Get to Play." Seattle Post-Intelligencer, 30 June 1980.

Riconoscimenti *Awards*

2004, Stewardship Award, Cultural Landscape Foundation.

2002, State Historic Registration, State of Washington - National and State Historic Preservation Board.

1999, Historic Landmark Status, City of Seattle - Landmark Board.

1997, Friends of GWP, Founded by Cheryl Trivison.

1993, Excellence on the Waterfront International Honor Award, Waterfront Center.

1981, President's Award of Design Excellence, American Society of Landscape Architects.

1980, Certificate of Design Excellence in Environmental Design, Print Casebooks 4.

1980, Special Mention Award for Project Design (Open Spaces), Urban Environmental Design National Awards.

1980, First Award, Adaptive Re-Use, Shoreline Design Awards.

1979, Honor Award, Washington Chapter ASLA.

1976, Award for Excellence of Realtors, Seattle-King County Board.

1975, Award for Excellence, Design and Environment.

Mostre *Exhibitions*

2003, University of Washington.

1998, Harvard School of Design.

1993, XIX Congress of International Union of Architecture (UIA Barcelona 96) "Present and Futures. Architecture in Cities features Gas Works Park as part of Terrain Vague", Barcelona.

BLOEDEL RESERVE

Bibliografia (parziale) *Bibliography (partial)*

Ament, Deloris Tarzan. "The Great Escape: Bainbridge Island's Bloedel Reserve Offers a Tranquil, Green Retreat to Soothe the Spirit." *Seattle Times* (9 October 1988) K1, K5.

American Academy in Rome, "Annual Exhibition 1998.", June 12-July 12, 1998 (exhibition catalog): 52-55, 106.

Appleton, Jay. The Experience of Landscape. Rev. ed. London: John Wiley and Sons, 1996: 248, 250, 254.

"Bloedel Reserve: Jury Comments." [ASLA President's Award of Design Excellence] (Architectural League of New York's Inhabited Landscapes exhibition), *Places* 4:4 (1987): 14-15.

Botta, Marina. "Seattle: 140 Acres of Green Rooms." Abitare (Italy) 272 (March 1989): 219-23.

Frankel, Felice (photographs) and Jory Johnson (text). "The Bloedel Reserve." *Modern Landscape Architecture: Redefining the Garden.* New York: Abbeville Press, 1991: 52-69.

Frey, Susan Rademacher, "A Series of Gardens." Landscape Architecture Magazine (September 1986): cover, 55-61, 128.

Haag, Richard. "Contemplations of Japanese Influence on the Bloedel Reserve." Washington Park Arboretum Bulletin 53:2 (Summer 1990): 16-19.

Haddad, Laura. "Richard Haag: Bloedel Reserve and Gas Works Park." Book Review. Arcade, 16.3 (Spring 1998): 34.

Illman, Deborah. 1998 UW Showcase: Arts and Humanities at the University of Washington. Calendar. Seattle: Litho Craft, 1998, March.

Judd, Ron C. "Bloedel Reserve Worth the Wait." *Seattle Times.* 13 May 1999, D7.

Kreisman, Lawrence. "Shapely Estate." Seattle Times Pacific Magazine, May 14, 1989: 22-23, 26-32.

Kreisman, Lawrence. "The Bloedel Reserve – Gardens in the Forest." The Arbor Fund of Bloedel Reserve. 1988.

Lockman, Heather. "Gardens: Nature in Gentle Custody." Architectural Digest (June 1984): 122-126.

Lowry, Deci. "The Moss Garden at Bloedel Reserve." *Pacific Horticulture* (Spring 1990): 16-20.

McGonigal, Steve. "Bloedel Reserve–Bainbridge Island." *Balls and Burlap* (Sumner, WA) (June 1985) 7.

Meyer, Elizabeth. "Terrible Beauty: The Sublime in Bloedel Reserve and Gasworks Park." *GSD News* (Cambridge) Fall 1996: 24-31.

Saunders, William S., ed. *Richard Haag: Bloedel Reserve and Gas Works Park.* Princeton Architectural Press and Harvard University Graduate School of Design, 1998.

Simpson, Nan Booth. "Inspired Intuition." Garden Design (May/June 1993): 42-47.

Spirn, Anne Whiston Spirn, "Perspectives: Essays on the State of Landscape Architecture." Progressive Architecture (August 1991): 93.

Streatfield, David. "The Resonance of Japan in Pacific Northwest Gardens." Washington Park Arboretum Bulletin. Vol 60. No. 1 (Winter 1998): 5.

Thompson, J. William. "Landscape of Dream: Warrior of Vision." Landscape Architecture (September 1989) 79: 7, 80-86.

Richard Haag: Conferenze, Mostre e Riconoscimenti

Richard Haag: Conferences, Exhibitions and Awards

2001, National Building Museum, "Spotlight on Design: Richard Haag", Washington, DC, April 12.

1998, American Academy in Rome, "Annual Exhibition 1998", June 12-July 12.

1996, Harvard University Graduate School of Design "Exploring the Landscape Architecture of Richard Haag", April/May.

1990, Istituto Nazionale Arredo Urbano E Strutture Ambientali, "Urban Green Areas: Woods and Ecological Parks", Verona e *and* Roma.

1987, The Architectural League of New York City, "The Inhabited Landscape".

1986, Award for Design Excellence, American Society of Landscape Architects President's.